Indeterminate Equations

Xing Zhou

Math for Gifted Students

http://www.mathallstar.com

 *use your mobile device to scan this QR code for more resources including books, practice problems, online courses, and blog.*

This book was produced using the LaTeX system.

Contents

CONTENTS

CONTENTS

Preface

Welcome to Math All Star© series!

Math All Star originates from a series of lectures given to a group of gifted middle school students with a love for mathematics and an interest in participating in competitions such as MathCounts, AMC, and AIME. These lectures aim to strengthen their problem-solving abilities and to introduce effective techniques that are not typically taught in the classroom.

As the popularity of Math All Star grew, the author began to upload lecture materials to create online courses, thereby providing students with the opportunity to progress at their own paces.

Since then, course materials have constantly been reviewed and updated to reflect student feedback and the observations made during lectures. Recent competition problems are also continuously analyzed and referenced to ensure the relevance of the contents. These course materials are the foundations of this Math All Star series.

Because competition math is a diversified subject that covers both a wide breadth and depth of topics, it is quite challenging to effectively cover all the material in one book that is appropriate for every interested student. Consequently, the author has decided to write a series of books, with each one focusing on a particular topic. Students are encouraged to pick and choose where to begin, depending on their individual skill levels and needs.

In addition to these books, the Math All Star website provides extra practice problems and serves as a highly recommended supplemental learning resource.

If there are any questions, comments, or concerns, please visit the website or email `contact@mathallstar.com`.

Happy learning!

 *To visit the Math All Star website, scan this QR code or go directly to*
http://www.mathallstar.com

Chapter 1

Introduction

Solving indeterminate equations is a popular subject in math competitions at all levels, from AMC 8 to IMO[1].

Despite its popularity, how to solve indeterminate equations is rarely discussed in classrooms. As a result, many students are lack of necessary knowledge and skills to tackle such problems. This book is to discuss various types of indeterminate equations and corresponding solving techniques. Upon completing this book, readers should be able to recognize and solve these indeterminate equations comfortably.

1.1   Indeterminate Equation Explained

Equations becomes indeterminate when the number of variables exceeds the number of given equations. Here is an example:

$$x + xy + y = 8 \tag{1.1}$$

---

[1]International Mathematical Olympiad. It is the most prestigious international math competition at high school level.

In this example, there are two variables, x and y, with just one equation given. Consequently, there are infinitely many pairs of real numbers (x, y) that can satisfy this equation.

With that said, we often are only interested in their integer solutions. Under this assumption, an indeterminate equation may have no solution, limited number of solutions, or infinitely many solutions. Therefore, the objective of solving an indeterminate equation is to:

- prove it has no solution, or

- list all the solutions, or

- find its general solution form[2]

If any of these three results is obtained, the corresponding indeterminate equation is said to have been solved.

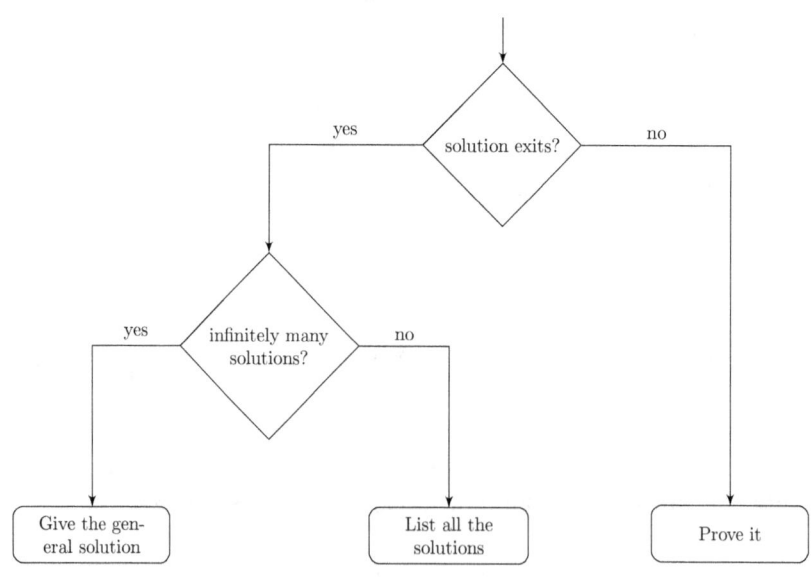

---

[2]A general solution is usually a parametric expression that can generate all the solutions to an indeterminate equation. It is especially useful when the given equation has infinitely many solutions. We will discuss it later in the book.

1.2   Conventions

In order to keep the contents concise and focused, the following conventions will be followed throughout this book, unless specially mentioned otherwise:

- Letters such as x, y, and z are variables that are to be solved

- Letters such as a, b, c, k, and n are known constants

- The solution is limited to integers

For example, $x^n + y^n = z^n$ is an equation with respect of x, y, and z, where n is a known quantity. Therefore it includes equations such as $x^2 + y^2 = z^2$, but not $2^x + 3^y = 4^z$. The latter is in the family of $a^x + b^y = c^z$.

1.3   Practice

Practice 1

What is an indeterminate equation?

Practice 2

What are the objectives when working on an indeterminate equation?

Practice 3

What do you think is the key to solve an indeterminate equation?

Chapter 2

The Factorization Method

Factorization is the most important elementary technique to solve an indeterminate equation. It is the method of choice in entry level competitions such as AMC8. At more advanced levels when problems become more challenging, factorization often serves as an critical stepping stone in complete solutions.

2.1 Solve Equation: $x + xy + y = n$

Many indeterminate equations can be categorized into several well known families. These families of equations usually have well-known solutions. For example, *Equation 1.1* given in *Chapter 1* is an instance of the equation family $x + xy + y = n$ where n is a known integer. Such equations can be solved using the factorization method.

There are two basic strategies to apply the factorization method. One is based on polynomial factorization, and the other relies on integer divisibility. Both strategies are illustrated in the following example.

Example 2.1.1

Find all ordered pairs of non-negative integers (x, y) that satisfy the equation $x + xy + y = 8$.

Solution 1

The given equation can be transformed into:

$$x + xy + y = 8$$
$$x + xy + y + 1 = 9$$
$$(x + 1)(y + 1) = 9$$

Because both x and y are integers, so will $(x+1)$ and $(y+1)$ be. It follows that $(x + 1)$ and $(y + 1)$ must be paired divisors of 9. As such, we can list all these paired divisors, and find corresponding x and y:

$x + 1$	$y + 1$	x	y
1	9	0	8
3	3	2	2
9	1	8	0

Hence, we conclude the given equation has three non-negative solutions: $(0, 8)$, $(2, 2)$, and $(8, 0)$.

Done.

Caution: It is important to note whether negative integer solutions are allowed or not.

If *Example 2.1.1* asks for integer solutions instead, numbers such as -1, -3, and -9 should be included as divisors of 9 as well.

The above solution is based on polynomial factorization. An alternative which utilizes integer divisibility is shown below.

Solution 2

Let's solve x with respect to y, i.e. to express x using y:

$$x + xy + y = 8$$
$$x\,(1+y) + y = 8$$
$$x = \frac{8-y}{1+y}$$
$$x = \frac{9-(1+y)}{1+y} \qquad (2.1)$$
$$x = \frac{9}{1+y} - 1 \qquad (2.2)$$

Given x is a non-negative integer, we find $\frac{9}{1+y}$ must be a positive integer. It follows that $(1+y)$ must be a positive divisor of 9. We then list all the positive integers that divide 9, and determine the corresponding y and x:

$1+y$	y	x
1	0	8
3	2	2
9	8	0

Done.

The two solutions give the same answer.

It is worth mentioning that step 2.1 is critical to make the 2nd solution work. The variable y has to appear only in the denominator in order to apply the divisibility argument. Otherwise we can only claim that $(1+y)$ is a divisor of $(8-y)$ which cannot be used directly[1].

Despite its simplicity, *Example 2.1.1* reveals two important aspects of solving an indeterminate equation.

The first one is the relevance of limiting the solutions to integers. Clearly, both solutions exploit this feature. In fact, all the solving methods utilize this in one way or another.

Secondly, the key to solve an indeterminate equation is to uncover intrinsic properties of and relationships among variables. For example, in *Example 2.1.1*, we find that both $(x+1)$ and $(y+1)$ are divisors of 9. This relationship is not obvious in the original expression, but plays a vital role in the final solution.

> The key to solve an indeterminate equation is to uncover intrinsic properties of and relationship among variables. Different solving methods are essentially different ways to reveal such information.

2.2 Solve Equation: $\frac{1}{x} + \frac{1}{y} = \frac{1}{n}$

Equations in the form of $\frac{1}{x} + \frac{1}{y} = \frac{1}{n}$ is another category that can be solved using the factorization method[2].

Here is an example:

---

[1]Having said that, it is possible to apply the inequality method to tackle the problem if it is not possible to have the numerator part a constant. The inequality method will be discussed in the next chapter.

[2]Please note that the left side has only two terms. If there are more terms, e.g. $\frac{1}{x} + \frac{1}{y} + \frac{1}{z} = \frac{1}{n}$, it cannot be solved by the factorization method alone. Solving such equations will be discussed in *Chapter 3*.

Example 2.2.1

Solve this equation in positive integers: $\frac{1}{x} + \frac{1}{y} = \frac{1}{3}$

Solution

Rewrite the equation as follow:

$$3x + 3y = xy$$
$$xy - 3x - 3y = 0$$
$$xy - 3x - 3y + 9 = 9$$
$$(x-3)(y-3) = 9 \qquad (2.3)$$

Because both x and y are integers, $(x-3)$ and $(y-3)$ will be integers too. Therefore, both of them must be paired divisors of 9. This leads to the following answers by applying a similar divisibility analysis as that is used in *Example 2.1.1* on *page 6*.

$$(x,y) = (4, 12), (6, 6), (12, 4)$$

Done.

💡 *Tip: Example 2.2.1 shows that the following relationship always holds:*

$$\frac{1}{x} + \frac{1}{y} = \frac{1}{n} \implies (x-n)(y-n) = n^2 \qquad (2.4)$$

It is therefore interesting to ask the following question:

Example 2.2.2

Given a positive integer n, how many positive integer solutions does the equation $\frac{1}{x} + \frac{1}{y} = \frac{1}{n}$ have?

Based on *Equation 2.4*, it is easy to derive that the answer is the number of positive divisors that n^2 has. Counting an integer's divisors is a basic counting question that has been discussed in the book *Counting*[3].

In a nutshell, if a positive integer n can be prime factorized into:

$$n = p_1^{a_1} \cdot p_2^{a_2} \cdots p_k^{a_k}$$

then n will have this many positive divisors:

$$(a_1 + 1)(a_2 + 2) \cdots (a_k + 1)$$

Noting $n^2 = p_1^{2a_1} \cdot p_2^{2a_2} \cdots p_k^{2a_k}$, the following statement holds:

Theorem 2.2.1

Let n be a positive integer whose prime factorization is

$$n = p_1^{a_1} p_2^{a_2} \cdots p_k^{a_k}$$

then the equation

$$\frac{1}{x} + \frac{1}{y} = \frac{1}{n}$$

has $(2a_1 + 1)(2a_2 + 1) \cdots (2a_k + 1)$ positive integer solutions.

2.3 Difference of Squares

Difference of squares, $x^2 - y^2 = (x + y)(x - y)$, is a popular polynomial factorization formula. It is also frequently used to solve indeterminate equations.

---

[3]Math for Gifted Students: Counting, author Xing Zhou, ISBN: 978-1506119151

When applying the difference of squares method, sometimes it is necessary to transform the equation to a proper form first.

Example 2.3.1

Find all integer pairs (x, y) that satisfy the equation:

$$(xy - 4)^2 = x^2 + y^2$$

The left side of this equation is a perfect square, but the right side is the sum of two squares. As a result, the difference of squares formula cannot be directly applied. However, it is possible to transform $x^2 + y^2$ into a perfect square if a $2xy$ term can be constructed.

This observation leads to the following solution.

Solution

The given equation is equivalent to:

$$(xy - 4)^2 = x^2 + y^2$$
$$((xy - 3) - 1)^2 = x^2 + y^2$$
$$(xy - 3)^2 - 2(xy - 3) + 1 = x^2 + y^2$$
$$(xy - 3)^2 - 2xy + 6 + 1 = x^2 + y^2$$
$$(xy - 3)^2 + 7 = (x + y)^2$$
$$(x + y)^2 - (xy - 3)^2 = 7$$
$$(x + y + xy - 3)(x + y - xy + 3) = 7$$

The last equation leads to the following systems:

$$\begin{cases} x + y + xy - 3 = \pm 1 \\ x + y - xy + 3 = \pm 7 \end{cases} \quad \text{or} \quad \begin{cases} x + y + xy - 3 = \pm 7 \\ x + y - xy + 3 = \pm 1 \end{cases}$$

Solving these systems leads to two non-negative integer solutions:

$$(x, y) = (0, \pm 4), (\pm 4, 0)$$

Done.

2.4   More Examples

While equations in the form of $x + xy + y = n$ or $\frac{1}{x} + \frac{1}{y} = \frac{1}{n}$ are easy to be factorized, some other equations may require careful handling. *Example 2.3.1* is such an example. Solid polynomial transformation and factorization skill plays a critical role when applying the factorization method.

Here is an example that utilizes variable substitution to help polynomial factorization.

Example 2.4.1

Solve the following equation in integers: $x^2(y-1)+y^2(x-1) = 1$.

(Polish Mathematical Olympiad)

Solution

Letting $x = u + 1$ and $y = v + 1$, the equation becomes:

$$(u + 1)^2 v + (v + 1)^2 u = 1$$
$$(u^2 + 2u + 1)v + (v^2 + 2v + 1)u = 1$$
$$uv(u + v) + 4uv + (u + v) = 1$$
$$uv(u + v + 4) + (u + v + 4) = 5$$
$$(u + v + 4)(uv + 1) = 5$$

Therefore one of the following relationships must hold:

$$\begin{cases} u + v + 4 & = \pm 1 \\ uv + 1 & = \pm 5 \end{cases} \quad \text{or} \quad \begin{cases} u + v + 4 & = \pm 5 \\ uv + 1 & = \pm 1 \end{cases}$$

Solving these systems leads to $(u, v) = (0, 1), (1, 0), (1, -6)$ and $(-6, 1)$. Accordingly, $(x, y) = (1, 2), (2, 1), (2, -5)$, and $(-5, 2)$.

Done.

Example 2.4.2

Solve the following equation in integers

$$x^3 + y^3 + z^3 = x + y + z = 3$$

This system has two equations with three variables. When $x^3 + y^3 + z^3$ and $x + y + z$ appear simultaneously, it is natural to relate them to polynomial identity based transformation.

Solution

The identity

$$(x + y + z)^3 = x^3 + y^3 + z^3 + 3(x + y)(y + z)(z + x)$$

gives us

$$
\begin{aligned}
& (x + y)(y + z)(z + x) \\
={} & ((x + y + z)^3 - (x^3 + y^3 + z^3)) \div 3 \\
={} & (3^3 - 3) \div 3 \\
={} & 8
\end{aligned}
$$

Because of the given condition $x + y + z = 3$, we know $(x + y)$, $(y + z)$ and $(z + x)$ on the left side can be replaced with $(3 - z)$, $(3 - x)$ and $(3 - y)$, respectively:

$$(3 - x)(3 - y)(3 - z) = 8 \tag{2.5}$$

On the other hand, it holds that:

$$(3 - x) + (3 - y) + (3 - z) = 9 - (x + y + z) = 6 \tag{2.6}$$

Equation 2.6 implies $(3 - x)$, $(3 - y)$, and $(3 - z)$ must be all even, or exactly one of them is even.

Case 1: all the three numbers are even

By *Equation 2.5*, the relationship $|3 - x| = |3 - y| = |3 - z| = 2$ must hold. This means x, y, z all equal 1 or 5. Because $x + y + z = 3$, the only possibility is $x = y = z = 1$.

Case 2: exactly one of them is even

In this case, the only possibility is:

- One of $|3 - x|$, $|3 - y|$, and $|3 - z|$ must equal 8, and

- The other two must equal 1

Let's say $|3 - x| = 8$ and $|3 - y| = |3 - z| = 1$, the only possible combination is $(x, y, z) = (-5, 4, 4)$ if we also consider $x + y + z = 3$.

Because any of $|3 - x|$, $|3 - y|$ and $|3 - z|$ can equal 8, therefore the solutions are the permutations of $(-5, 4, 4)$.

In conclusion, the given equation has four integer solutions

$$(x, y, z) = (1, 1, 1), (-5, 4, 4), (4, -5, 4), (4, 4, -5)$$

Done.

2.5   Practice

Practice 1

Solve this equation in integers $3(x + y) = xy + 8$.

Practice 2

A line $y = px$, where p is a non-zero integer, intersects line $y = x + 10$ at a grid point whose x and y coordinates are both integers. How many such lines $y = px$ exist?

Practice 3

How many ordered integer pairs (x, y) are there that can satisfy the following equation?

$$\frac{x+3}{x+1} - y = 0$$

Practice 4

A grid point is defined as a point whose x and y coordinates are both integers. How many grid points will the function plot

$$y = \frac{x+12}{2x-1}$$

pass?

Practice 5

Solve the following equation in positive integers:

$$\frac{1}{x} + \frac{1}{y} = \frac{1}{5}$$

Practice 6

For each positive integer n, let $s(n)$ denote the number of ordered positive integer pair (x, y) for which

$$\frac{1}{x} + \frac{1}{y} = \frac{1}{n}$$

Find all positive integers n for which $s(n) = 5$.

(Indian Mathematical Olympiad)

Practice 7

Find all integer pairs (x, y) such that $x - y^4 = 4$, where x is a prime number.

Practice 8

Let p and q be two distinct prime numbers. Solve the following equation in positive integers:

$$\frac{p}{x} + \frac{q}{y} = 1$$

Practice 9

For any given positive integer $n > 2$, show that there exists a right triangle satisfying the following conditions:

- the lengths of all its three sides are integers, and

- one of these lengths equals n

Chapter 3

The Inequality Method

Conceptually, one variable is one moving part. If there are too many moving parts, it will be challenging to fix all of them at once. The essence of the inequality method is to first determine the range of one variable, thus to reduce the number of moving parts.

3.1 Solve Equation $\frac{1}{x} + \frac{1}{y} = \frac{m}{n}$

An equation in the form of $\frac{1}{x} + \frac{1}{y} = \frac{m}{n}$ looks similar to $\frac{1}{x} + \frac{1}{y} = \frac{1}{n}$. While the latter can always be solved using the factorization method as discussed in *Section 2.2*, the former may or may not be solvable using polynomial transformation. This is because factorizing an equation in the form of $\frac{1}{x} + \frac{1}{y} = \frac{m}{n}$ may not be straightforward. In such cases, the inequality method comes handy.

Example 3.1.1

Solve the following equation in positive integers: $\dfrac{1}{x} + \dfrac{1}{y} = \dfrac{5}{6}$.

Solution

The given equation is equivalent to

$$6x - 5xy + 6y = 0 \tag{3.1}$$

Solve x with respect to y:

$$x = \frac{6y}{5y - 6} = 1 + \frac{y + 6}{5y - 6} \tag{3.2}$$

Clearly, it is not possible to transform the right side of this equation to a sum of an integer and a fraction whose numerator is a constant. As a result, the divisibility argument which is used in the 2nd solution of *Example 2.2.1* cannot be applied here.

However, it is perfectly acceptable to claim the following inequality relationship:

$$|5y - 6| \leq y + 6$$

Also considering *Equation 3.2* leads to:

$$0 < 5y - 6 \leq y + 6$$

or $y = 2, 3$. Setting these values to relationship 3.2 yields the following qualified solutions:

$$(x, y) = (2, 3), (3, 2)$$

Done.

3.2 Symmetrical Equations

Many indeterminate equations are symmetrical. Exploring the feature of symmetry often plays an important role when applying the inequality method.

Definition 3.2.1   Symmetrical Equation

Being symmetrical means exchanging variables in an equation does not change the nature of that equation at all.

For example, the equation $x + xy + y = 8$ in *Example 2.2.1* on *page 9* is symmetrical. This is because exchanging x and y leads to the following equation which is the same as the original one

$$y + yx + x = 8$$

Similarly, the equation $\frac{1}{x} + \frac{1}{y} = \frac{5}{6}$ in *Example 3.1.1* is also symmetrical.

A symmetrical equation enjoys the following property:

Theorem 3.2.1   Symmetrical Solutions

If $(a, b, \cdots, k)$ is one solution to a symmetrical equation, then any distinct permutation of $(a, b, \cdots, k)$ is a solution to that equation too.

Theorem 3.2.1 has a useful implication. When an equation is symmetrical, it is possible to assume an order of all the variables without loss of generality, e.g. $x \geq y \geq z \geq \cdots$. This is because if this equation has solutions, it is aways possible to swap relevant variables in order to satisfy this inequality relationship.

Upon having obtained all the solutions under the assumption of $x \leq y \leq z \leq \cdots$, all the solutions without such assumption are just distinct permutations of that set.

Such inequality assumption offers two benefits. The first is to reduce the number of scenarios to analyze. The second is to provide an additional condition upon which it can be easier to apply inequality analysis.

Let's re-examine *Example 3.1.1*.

Solve this equation in positive integers: $\dfrac{1}{x} + \dfrac{1}{y} = \dfrac{5}{6}$.

Alternative Solution

By symmetry, let's assume $x \leq y$. Hence

$$\frac{1}{x} \geq \frac{1}{y}$$

It follows that,

$$\frac{1}{x} \geq \frac{1}{2} \times \frac{5}{6} = \frac{5}{12} \tag{3.3}$$

or $x \leq 2$.

Testing $x = 1, 2$ respectively finds $(2, 3)$ is one solution. Therefore all the solutions to the given equations are

$$(x, y) = (2, 3), (3, 2)$$

Done.

This answer agrees with the previous solution. Note that *Equation 3.3* can also be explained by the fact that, as the larger of the two terms, $\frac{1}{x}$ must be at least equal to the average of the sum, i.e. $\frac{1}{2} \times \frac{5}{6}$.

> 💡 *Tip: This explanation is useful to simplify the inequality argument, especially when there are multiple variables involved.*

Despite simplicity, this example demonstrates the power of symmetry. It is especially useful when using the inequality method to solve symmetrical equations involving multiple variables.

💡 *Tip: When dealing with symmetrical equations, the following abbreviation appears frequently:*

WLOG = Without Loss Of Generality

3.3 Solve Equation: $\frac{1}{x} + \frac{1}{y} + \frac{1}{z} = \frac{1}{n}$

The first question to consider is whether the following two equations are the same type, thus can be solved using the same approach or not?

- $\frac{1}{x} + \frac{1}{y} = \frac{1}{n}$

- $\frac{1}{x} + \frac{1}{y} + \frac{1}{z} = \frac{1}{n}$

⏸ *Pause: think about it before continuing ...*

The answer is no. Even though they look similar, the 2nd question cannot be solved using the factorization method alone.

💡 *Tip: Seemingly subtle difference may be significant.*

The reason that the 2nd equation cannot be solved by factorization is that there are 3 variables now. If we solve x with respect to y and z, there are still two remaining variables which will prevent the divisibility argument from being applied in a straightforward way.

In order to solve the equation $\frac{1}{x} + \frac{1}{y} + \frac{1}{z} = \frac{1}{n}$, we turned to the inequality method.

Example 3.3.1

Solve the following equation in positive integers

$$\frac{1}{x} + \frac{1}{y} + \frac{1}{z} = \frac{3}{5}$$

(Romanian Mathematical Olympiad)

Solution

By the symmetrical argument, let's assume $0 < x \le y \le z$. It follows:

$$\frac{1}{x} < \frac{1}{x} + \frac{1}{y} + \frac{1}{z} \le \frac{3}{x}$$

Then $\dfrac{1}{x} + \dfrac{1}{y} + \dfrac{1}{z} = \dfrac{3}{5} \implies \dfrac{1}{x} < \dfrac{3}{5} \le \dfrac{3}{x} \implies 2 \le x \le 5.$

Now we proceed with casework:

If $x = 2$, then $\dfrac{1}{y} + \dfrac{1}{z} = \dfrac{3}{5} - \dfrac{1}{2} = \dfrac{1}{10}.$

If $x = 3$, then $\dfrac{1}{y} + \dfrac{1}{z} = \dfrac{3}{5} - \dfrac{1}{3} = \dfrac{4}{15}.$

If $x = 4$, then $\dfrac{1}{y} + \dfrac{1}{z} = \dfrac{3}{5} - \dfrac{1}{4} = \dfrac{7}{20}.$

If $x = 5$, then $\dfrac{1}{y} + \dfrac{1}{z} = \dfrac{3}{5} - \dfrac{1}{5} = \dfrac{2}{5}.$

The equivalent equation in every case is in the form of:

$$\frac{1}{x} + \frac{1}{y} = \frac{1}{n} \quad \text{or} \quad \frac{1}{x} + \frac{1}{y} = \frac{m}{n}$$

They can all be solved by using the relevant techniques that have been discussed in *Section 2.2* and *Section 3.1*, respectively. Solving these equations leads to the following solutions under the assumption $0 < x \le y \le z$.

(2, 11, 110), (2, 12, 60), (2, 14, 35), (2, 15, 30), (2, 20, 20), (3, 4, 60), (3, 5, 15), (3, 6, 10), (4, 4, 10), and (5, 5, 5).

Therefore, all the solutions are just distinct permutations of the above set.

Done.

This example shows how the symmetrical argument helps to identify a special variable for analysis. In this case, it is the smallest among all the variables, i.e. x. In other cases, it can be the largest one or with some special properties, such as odd-even parity and so on. Regardlessly, the goal is to find the candidate that is to be tackled first. Then, the inequality method determines the possible values of this variable. Afterwards, by applying the casework, the original equation with three variables is simplified to several equations with just two variables. Such approach is typical in employing the inequality method.

3.4   Sum of Squares

Sum of squares is a frequently appeared pattern. The technique is to transform an indeterminate equation into the following form:

$$(\cdots)^2 + (\cdots)^2 + \cdots + (\cdots)^2 = n$$

where n is a non-negative constant, and every term on the left side contains one or more variables.

It is obvious that the following must hold for every term on the left side:

$$0 \le (\cdots)^2 \le n$$

Therefore the original problem becomes a system of equations which is easier to solve.

In particular, if $n = 0$, then all the terms must equal 0. Let's review an example:

Example 3.4.1

Solve the following equation in integers the equation:

$$x^2 + 2x + y^2 - 4y = -5$$

Solution

The given equation is equivalent to:

$$(x+1)^2 + (y-2)^2 = 0$$

Therefore the following must hold:

$$\begin{cases} x + 1 &= 0 \\ y - 2 &= 0 \end{cases}$$

This leads to a unique solution $(-1, 2)$.

Done.

Transformation used in *Example 3.4.1* is relatively straightforward. Sometime it may be a bit more complex. In such cases, it is helpful to find obvious solutions first, and to tackle more challenging solutions next.

Let's consider the following example:

Example 3.4.2

Find all integer solutions to the equation $x^3 + y^3 = (x + y)^2$.

Solution

It is clear that when $x = -y$, the equation is an identity. Therefore all the integer pairs $(k, -k)$ are solutions to the given equation.

When $x + y \neq 0$, dividing $(x + y)$ on both sides leads to

$$x^2 - xy + y^2 = x + y$$

This equation can be re-written as:

$$x^2 - xy + y^2 - x - y = 0 \tag{3.4}$$

or

$$(x - y)^2 + (x - 1)^2 + (y - 1)^2 = 2 \tag{3.5}$$

It follows that two of the three terms must equal 1, and the other term equals 0. There are totally five ordered integer pairs that meet this condition: $(0, 1)$, $(1, 0)$, $(1, 2)$, $(2, 1)$, and $(2, 2)$.

Hence the answers are:

- $(k, -k)$ where k is any integer, and

- $(0, 1), (1, 0), (1, 2), (2, 1)$, and $(2, 2)$

Done.

In this example, transforming *Equation 3.4* to 3.5 may not appear apparent. An alternative solution will be discussed in *Chapter 4 The Quadratic Method*.

💡 *Tip: Being able to solve one problem using multiple techniques is certainly beneficial.*

3.5   The Squeeze Technique

Squeeze is another technique that is built upon the inequality principle. In order to understand this technique, let's first investigate the following example.

Example 3.5.1

Solve this equation in positive integers: $y^2 = x^2 + x + 1$.

This problem is equivalent to finding a perfect square that can be expressed as $x^2 + x + 1$. Is it possible?

Solution

It is obvious that $x > 0 \implies x^2 < y^2 = x^2 + x + 1 < (x+1)^2$.

This implies y^2 is between x^2 and $(x+1)^2$. Note that x and $(x+1)$ are two consecutive integers. Hence it is impossible to have another integer whose square is between squares of two consecutive integers.

Done.

Example 3.5.1 shows that the essence of the squeeze method is to find an upper boundary and a lower boundary of the given expression. Hence, the critical step is to approximate a polynomial of many terms using a square (or a cube and so on, as necessary). Such approximations provide the needed boundaries to squeeze the target variables.

Here is another example:

Example 3.5.2

Solve the following equation in integer

$$x^3 + (x+1)^3 + (x+2)^3 + \cdots (x+7)^3 = y^3$$

(Hungarian Mathematical Olympiad)

Though it is not difficult to expand and then consolidate the terms on the left side, it is still a challenge to solve a cubic equation. However, as both sides are cubic, it may be possible to determine the boundaries of y using the squeeze method.

Solution

Let $P(x) = x^3 + (x+1)^3 + (x+2)^3 + \cdots + (x+7)^3$, then

$$\begin{aligned} P(x) \ &= 8x^3 + 84x^2 + 420x + 784 \\ &= (2x+7)^3 + 63(2x+7) \\ &= (2x+10)^3 - 36(x+2)(x+3) \end{aligned}$$

If $x \geq 0$ then $(2x+7)^3 < P(x) < (2x+10)^3$. This means

$$(2x+7)^3 < y^3 < (2x+10)^3$$

i.e.
$$y = 2x+8, y = 2x+9$$

Do casework:

- $y = 2x + 8 \implies 8x^3 + 84x^2 + 420x + 784 = (2x+8)^3$

- $y = 2x + 9 \implies 8x^3 + 84x^2 + 420x + 784 = (2x+9)^3$

Both can be simplified to standard quadratic equations. Neither has integer solution. Or, in another word, this equation does not have non-negative solution.

If $x < 0$, we note that if (x, y) is a solution, so will $(-x - 7, -y)$ be.[1] Hence, it will not be solvable for $x \le -7$ because otherwise it will contradict the previous conclusion of no non-negative solution.

Hence, all the candidate solutions must satisfy $-7 < x < 0$. Substituting x with $-1, -2, \cdots, -6$, respectively, finds the following four solutions: (-2,6), (-3, 4), (-4, -4), and (-5, -6).

Done.

3.6　Practice

Practice 1

Solve the following equation in integers: $y^2 = (x + 1)(x + 2)$.

Practice 2

Solve in positive integers the equation: $\frac{1}{x} + \frac{1}{y} + \frac{1}{z} = 1$.

Practice 3

Find all positive integer triples (x, y, z) such that

$$3(xy + yz + zx) = 4xyz$$

---

[1] In a more general sense, polynomial $x^3 + (x + 1)^3 + (x + 2)^3 + \cdots + (x + 7)^3$ is symmetrical with respect to $x + 4$, its middle term.

Practice 4

Solve the following equation in integers:

$$x + \frac{1}{y + \frac{1}{z}} = \frac{10}{7}$$

Practice 5

Solve the following equation in positive integers

$$\left(1 + \frac{1}{x}\right)\left(1 + \frac{1}{y}\right)\left(1 + \frac{1}{z}\right) = 2$$

(UK Mathematical Olympiad)

Practice 6

Let integers a, b, and c satisfy $a - 2b = 4$ and $ab + c^2 - 1 = 0$. Find the value of $a + b + c$.

Practice 7

Find all positive integers n and k_i $(1 \leq i \leq n)$ such that

$$k_1 + k_2 + \cdots + k_n = 5n - 4$$

and

$$\frac{1}{k_1} + \frac{1}{k_2} + \cdots + \frac{1}{k_n} = 1$$

(Putnam Mathematical Competition)

Practice 8

Solve the following equation in positive integers:

$$xy + yz + zx - xyz = 2$$

Practice 9

If integers x, y, and z are all greater than 2. Solve the following equation:

$$\frac{1}{x} + \frac{1}{y} - \frac{1}{z} = \frac{1}{2}$$

Practice 10

Find all integers x that can satisfy the equation:

$$\frac{1}{x} + \frac{1}{x+1} + \frac{1}{x+2} = \frac{13}{12}$$

Chapter 4

The Quadratic Method

Like many other mathematical problems, an indeterminate equation may be solved in multiple ways. The quadratic method may come handy when the target equation can be organized as a quadratic equation with respect to one variable.

4.1 The Principle

For a quadratic equation to be solvable, its discriminant

$$\Delta = b^2 - 4ac$$

must be non-negative. Furthermore, by the quadratic formula

$$x_{1,2} = \frac{-b \pm \sqrt{b^2 - 4ac}}{2a}$$

if the roots are integers, then

(i) its $\Delta = b^2 - 4ac$ must be a square number, and

(ii) $-b \pm \sqrt{b^2 - 4ac}$ must be divisible by $2a$

Condition (i) above often leads to a new indeterminate equation:

$$b^2 - 4ac = k^2 \qquad (4.1)$$

where k is an integer.

Equation 4.1 usually can be solved by the factorization method because it is a natural candidate for the difference of square technique:

$$b^2 - k^2 = 4ac$$

Condition (ii) above provides an opportunity to reveal insights of k using divisibility analysis. For example, if k has to be odd as a result, then *Equation 4.1* can be rewritten as

$$b^2 - 4ac = (2m + 1)^2$$

4.2 Examples

The first step to apply this quadratic method is to re-organize an indeterminate equation to a quadratic equivalent with respect of one variable.

Let's illustrate this by revisiting *Example 3.4.1* on *page 24*.

Example 4.2.1

Solve in integers the equation $x^2 + 2x + y^2 - 4y = -5$.

The given equation has two variables and the highest power is 2. Therefore it can be an ideal candidate to apply the quadratic method.

The solution given below is to organize it as a quadratic equation with respect to x. It is also possible to organize it with respect to y. In most cases, the choice to choose x or y depends on which can lead to a simpler discriminant expression. In this example, choosing x or y does not matter.

Solution

The given equation can be organized as a quadratic equation with respect to x:

$$x^2 + 2x + (y^2 - 4y + 5) = 0 \qquad (4.2)$$

In order for *Equation 4.2* to have at least one integer solution, its discriminant must be non-negative:

$$\Delta = 2^2 - 4 \times (y^2 - 4y + 5) \geq 0$$

This is equivalent to $(y - 2)^2 \leq 0$, which implies $y = 2$. Setting $y = 2$ in the original equation yields $x = -1$.

Therefore, the given equation has only one solution: (-1, 2).

Done.

Sometimes, it may be necessary to construct a quadratic equation before applying this technique.

Example 4.2.2

There exist some positive integers x and y such that

$$\frac{x}{y} + \frac{15y}{4x}$$

is an integer. Find all such (x, y) pairs where x and y are relatively prime.

The expression is not quadratic. However, given both $\frac{x}{y}$ and $\frac{y}{x}$ appearing simultaneously, this expression can be transformed into a quadratic one by a frequently used substitution technique.

Solution

Let $u = \frac{x}{y}$, then the given problem is equivalent to:

$$u + \frac{15}{4u} = k$$

where k is an integer. It is obvious that u is a positive rational number because both x and y are positive integers.

Rewriting this relationship leads to:

$$4u^2 - 4ku + 15 = 0 \tag{4.3}$$

Because *Equation 4.3* is solvable in rational number, its discriminant must be a square number. Let

$$\Delta = 16k^2 - 4 \times 4 \times 15 = n^2$$

where n is an integer. Or:

$$16k^2 - n^2 = 240$$

Clearly, n^2 is a multiple of $16 = 4^2$, setting $n = 4m$ leads to:

$$16k^2 - 16m^2 = 240$$
$$k^2 - m^2 = 15$$
$$(k + m)(k - m) = 15 \tag{4.4}$$

Equation 4.4 can be solved by the factorization method. Because both k and m are positive integers, we have $k + m > k - m$. Consequently, one of the two systems must hold:

$$\begin{cases} k + m &= 15 \\ k - m &= 1 \end{cases} \quad \text{or} \quad \begin{cases} k + m &= 5 \\ k - m &= 3 \end{cases}$$

Solving the above two systems leads to

$$(k, m) = (8, 7), (4, 1)$$

Setting $k = 8$ to the quadratic formula of *Equation 4.3*:

$$u = \frac{4 \times 8 \pm \sqrt{(4 \times 8)^2 - 4 \times 4 \times 15}}{2 \times 4} = 4 \pm \frac{7}{2} = \frac{15}{2}, \frac{1}{2}$$

Setting $k = 4$ leads:

$$u = \frac{4 \times 4 \pm \sqrt{(4 \times 4)^2 - 4 \times 4 \times 15}}{2 \times 4} = 2 \pm \frac{1}{2} = \frac{5}{2}, \frac{3}{2}$$

Therefore, we conclude there are four solutions:

$$(x, y) = (15, 2), (1, 2), (5, 2) \text{ and } (3, 2)$$

Done.

4.3   Practice

Practice 1

Comparing with other techniques such as the factorization method, what are benefits of the quadratic method?

Practice 2

Solve this equation in integers: $y^2 - x^2 - 3x = 5$.

Practice 3

Find all integer solutions to the following equation:

$$x^2 + 4xy + 5y^2 + 2x + 4y - 7 = 0$$

Practice 4

Solve the following equation in integers:

$$x^2 + xy + y^2 = 1$$

Practice 5

Solve this equation in integers: $x^3 + y^3 = (x + y)^2$.

Practice 6

Use the quadratic method to find all the integer pairs (x, y) that satisfy the equation:

$$(xy - 4)^2 = x^2 + y^2$$

Chapter 5

The Euclidean Method

The Euclidean method is a classical way to find the greatest common divisor of two integers. It is often taught in classrooms. In this chapter, we extend this method to solve certain indeterminate equations. As a closely related topic, Bézout's identity is also introduced and discussed.

5.1 Solve Equation: $ax + by = 1$

Equations in the form of $ax + by = 1$ is an important family of indeterminate equations. It is also closely related to the Bézout's identity which is an important topic in number theory.

First, let's examine the necessary condition for such an equation to be solvable.

Example 5.1.1

Solve the following equation in integers: $12x + 15y = 1$.

⏸ *Pause: think about it before continuing ...*

Solution

The given equation does not have any integer solution. This is because the left side is a multiple of 3, but the right side is not.

Done.

Despite its simplicity, *Example 5.1.1* reveals an important conclusion which can be formalized in the following way:

Theorem 5.1.1   Bézout's Identity

The equation $ax + by = 1$, where both a and b are integers, is solvable in integers if and only if a and b are relatively prime.

The logic presented in *Example 5.1.1* shows that if a and b are not relatively prime, then $ax + by = 1$ will be unsolvable in integers. In fact, it can be proved that if a and b are relatively prime, the equation $ax + by = 1$ will be solvable in integers.

5.2   Bézout's Identity

Theorem 5.1.1 presented in the previous section essentially states that whether a and b are relatively prime is equivalent to whether the equation $ax + by = 1$ is solvable in integers. We can either assert that the equation is solvable if a and b are relatively prime, or prove a and b are relatively prime by obtaining a pair of integers (x, y) to show this equation is solvable.

a and b are relatively prime	$\Longleftrightarrow$	$ax + by = 1$ is solvable in integers

Bézout's Identity has wide applications in number theory. Here, we just present one example.

Example 5.2.1

Show that n and $(n + 1)$ are relatively prime for any positive integer n.

This conclusion can be proved using the basic definition of relatively prime.

Solution 1

Let d be the greatest common divisor of n and $(n+1)$. Our goal is to prove $d = 1$.

As d is a divisor of both n and $(n + 1)$, let

$$\begin{cases} n & = d \cdot k \\ n + 1 & = d \cdot l \end{cases}$$

where k and l are positive integers satisfying $l > k$.

Subtracting the 1st equation from the 2nd yields:

$$1 = d \cdot (l - k)$$

Because both d and $(l - k)$ are positive integers, both of them must equal 1 in order to make their product 1. This means $d = 1$.

Done.

The above proof is correct. However, this problem can be proved more concisely by using the Bézout's identity.

Solution 2

In order to prove two integers, n and $(n+1)$, are relatively prime, it is sufficient to find two integers, x and y, such that

$$nx + (n+1)y = 1$$

Clearly, setting $x = -1$ and $y = 1$ satisfies the requirement.

Done.

5.3 Euclidean Method

The Euclidean method, sometime also referred as the Euclidean algorithm, is a classical way to find the greatest common divisor of two given integers. It also provides a guaranteed solution to solve an equation in the form of $ax + by = 1$ where a and b are two relatively prime integers.

This method is illustrated in the following example.

Example 5.3.1

Find one integer solution to the equation $41x + 17y = 1$.

Practically speaking, many simple equations in this form can be solved using the guessing-and-checking method. For example: $4x + 3y = 1$ has one solution $(-2, 3)$. With that said, when a and b are more complex which makes guessing-and-checking a challenging task, the Euclidean method can always be used to find one solution.

Solution

The Euclidean method has two steps.

Step 1: This step is exactly the same as to find the greatest common divisor of a and b. If the two given numbers are relatively prime, the final result will always be 1.

The process starts by dividing the larger one of a and b by the smaller one. It is then followed by repeatedly dividing the remainder obtained by previous step by the smaller one of the two previous numbers, until the remainder becomes 1.

$$41 \div 17 = 2 \dots R \quad 7 \qquad\qquad (41, 17)$$
$$17 \div 7 = 2 \dots R \quad 3 \qquad\qquad (17, 7)$$
$$7 \div 3 = 2 \dots R \quad 1 \qquad\qquad (7, 3)$$

Step 2: This step works bottom-up based on the equations obtained in step 1. In each iteration, the remainder is replaced by the relationship obtained earlier.

$$
\begin{aligned}
1 &= 7 - 3 \times 2 & replace \quad 1 \\
&= 7 - (17 - 7 \times 2) \times 2 & replace \quad 3 \\
&= 17 \times (-2) + 7 \times 5 & rearrange \\
&= 17 \times (-2) + (41 - 17 \times 2) \times 5 & replace \quad 7 \\
&= 41 \times 5 + 7 \times (-12) & rearrange
\end{aligned}
$$

The last relationship gives one solution to the original equation, i.e. $(x, y) = (5, -12)$.

Done.

5.4   The MOD Equation Method

While the Euclidean method is the most popular method to solve the equation $ax + by = c$, such equation can be solved using an alternative solution. In essence, it is an extension to divisibility based solutions.

Let's consider the following example:

Example 5.4.1

Find one integer solution to $11x + 7y = 1$.

Solution

The given equation is equivalent to:

$$y = \frac{1 - 11x}{7}$$

This implies $7 \mid (1 - 11x)$. or

$$1 - 11x \equiv 0 \pmod 7$$
$$11x \equiv 1 \pmod 7 \tag{5.1}$$

Equation 5.1 is a typical modular equation. Solving modular equations is an important topic in number theory. It will be covered in the number theory book. A straightforward approach is to test all the numbers from 0 to 6, which are all the possible remainders of (mod 7). Because 11 and 7 are relatively prime, one of these values will satisfy this equation.

In this case, setting $x = 2$ satisfies *Equation 5.1*. Accordingly $y = -3$.

Therefore, one solution to the equation $11x + 7y = 1$ is

$$(x, y) = (2, -3)$$

Done.

The MOD equation method comes handy when the coefficients, i.e. a and b, are relatively small. If they are large, it may be a time consuming process to test all the possible remainders.

5.5 Practice

Practice 1

What is the necessary condition for the equation

$$ax + by = 1$$

to be solvable in integers, where both a and b are integers?

Practice 2

What do you think is the necessary condition for the equation

$$ax + by = c$$

to be solvable in integers? Here, a, b and c are all integers, and c may or may not equal 1.

Practice 3

Create an equation in the form of $ax + by = 1$ yourself, and try to solve it.

Practice 4

Show that for any given positive integer n, the fraction

$$\frac{7n + 1}{14n + 3}$$

must be in its simplest form.

Chapter 6

General Solution

As mentioned in *Chapter 1 Introduction*, one needs to achieve one of the three objectives below in order to solve an indeterminate equation:

- Prove the equation has no solution, or

- List all its solutions, or

- Obtain its general solution

This chapter discusses what is a general solution, and how to obtain it.

6.1 Special v.s. General Solution

In order to understand the difference between a special solution and a general solution, let's revisit *Example 5.3.1*:

Find one integer solution to $41x + 17y = 1$.

The answer given in *Section 5.3* on *page 40* is $(5, -12)$. This answer is certainly correct. However, it is easy to verify that other integer pairs such as $(-12, 29)$ or $(22, -53)$ satisfy this equation too. In fact, it can be shown that this equation has infinitely many integer solutions. Hence, a question arises: how to describe all these solutions?

This can be achieved by using parametric equations. A parametric equation is an expression that contains one or more parameters. For instance, the following parametric equations can describe all the solutions to the equation $41x + 17y = 1$:

$$\begin{cases} x &= 5 - 17t \\ y &= -12 + 41t \end{cases} \tag{6.1}$$

where t is an integer.

When $t = 0$, *Equation 6.1* gives $(x, y) = (5, -12)$ which is the original solution provided in *Section 5.3*. The two additional solutions $(-12, 29)$ and $(22, -53)$ can be obtained by setting $t = 1$ and -1, respectively. It is easy to verify that setting t to any integer will produce one valid solution. Furthermore, it can be shown that all the solutions can be obtained by setting t to appropriate integers.

Such a parametric equations is called a general solution. A pair of integers (x, y) that satisfies the equation is called a special solution.

While special solutions to a given equation can be different, all the general solutions must be equivalent even though their expressions can be different. It is then preferable to give the simplest form as the answer.

For example, the 1st problem in 2015 USAMO is an indeterminate equation which has unlimited number of solutions. Its general solution can be written as:

$$\begin{cases} x & = \dfrac{1}{8} \times (3 \times (k^2 - 1) \pm k \cdot (k^2 - 9)) \\[2mm] y & = \dfrac{1}{8} \times (3 \times (k^2 - 1) \mp k \cdot (k^2 - 9)) \end{cases} \qquad (6.2)$$

where integer k is an odd integer.

This answer is correct. However *Equation 6.2* can be simplified to the following equivalent by setting $k = 2n + 1$:

$$\begin{cases} x & = n^3 + 3n^2 - 1 \\[2mm] y & = -n^3 + 3n + 1 \end{cases} \qquad (6.3)$$

where n is an integer parameter.

Clearly, *Equation 6.3* is simpler than 6.2, and thus is preferable.

6.2　General Solution to $ax + by = 1$

When an equation has infinitely many solutions, one need to obtain its general solution in order to claim victory. Many well-known equations have well-known general solutions. The equation $ax + by = 1$ is one of them.

The general solution of $ax + by = 1$ is given by:

$$\begin{cases} x & = x_0 - b \cdot t \\ y & = y_0 + a \cdot t \end{cases} \qquad (6.4)$$

where (x_0, y_0) is a special solution, and t is an integer parameter.

As stated earlier, different special solutions (x_0, y_0) may lead to seemingly different general solutions. However, all the general solutions to the same indeterminate equation are all equivalent. This is illustrated in the following example.

Example 6.2.1

Solve in integers the equation: $41x + 17y = 1$.

Letting $(x_0, y_0) = (5, -12)$ leads to the following general solution:

$$\begin{cases} x &= 5 - 17t \\ y &= -12 + 41t \end{cases} \tag{6.5}$$

where t is an integer.

Letting $(x_0, y_0) = (-12, 29)$ leads to the following general solution:

$$\begin{cases} x &= -12 - 17u \\ y &= 29 + 41u \end{cases} \tag{6.6}$$

where u is an integer[1].

To see *Equation 6.5* and 6.6 are equivalent, let's rewrite 6.6 as the following:

$$\begin{cases} x &= 5 - 17 - 17u &= 5 - 17(u + 1) \\ y &= -12 + 41 + 41u &= -12 + 41(u + 1) \end{cases} \tag{6.7}$$

Comparing *Equation 6.5* and 6.7, it is easy to see they are the same if we set $t = u + 1$. Because both t and u can take any integer value, t and $(u + 1)$ have a one-to-one mapping relationship. Therefore we conclude *Equation 6.5* and 6.7 are equivalent.

---

[1]Note that we can use any letter to represent a parameter. The letter u used in this case is to distinguish it from the letter t used in *Equation 6.5* in order to avoid unnecessary confusion.

6.3 Solving $ax + by = c$

Based upon the knowledge of equation $ax + by = 1$, it is natural to investigate the following equation:

$$ax + by = c \qquad (6.8)$$

where c is an integer which may or may not equal 1.

Employing a similar argument as that used in *Example 5.1.1* on *page 37*, it is easy to see that *Equation 6.8* is solvable in integers if and only if $\gcd(a, b) \mid c$. Furthermore, it can be shown that when this equation is solvable, its general solution is given by:

$$\begin{cases} x = x_0 - \dfrac{b}{\gcd(a, b)} \cdot t \\[3mm] y = y_0 + \dfrac{a}{\gcd(a, b)} \cdot t \end{cases} \qquad (6.9)$$

where (x_0, y_0) is any special solution, and t is an integer parameter.

Let's review an example.

Example 6.3.1

Solve the following equation in integers: $2015x + 299y = 117$.

Solution

Because $\gcd(2015, 299) = 13$ and $13 \mid 117$, this equation is solvable in integer.

The 1st step is to simplify the equation by dividing 13 on both sides:

$$155x + 23y = 9 \qquad (6.10)$$

The 2nd step is to find a special solution to *Equation 6.10*. This can be done by guessing-and-checking, the Euclidean method, or the MOD equation method. Here, the Euclidan method is used.

$$155 = 23 \times 6 + 17$$
$$23 = 17 \times 1 + 6$$
$$17 = 6 \times 2 + 5$$
$$6 = 5 \times 1 + 1$$

This is followed by:

$$
\begin{aligned}
1 &= 6 - 5 \times 1 \\
&= 6 - (17 - 6 \times 2) \times 1 \\
&= 17 \times (-1) + 6 \times 3 \\
&= 17 \times (-1) + (23 - 17 \times 1) \times 3 \\
&= 17 \times (-4) + 23 \times 3 \\
&= (155 - 23 \times 6) \times (-4) + 23 \times 3 \\
&= 155 \times (-4) + 23 \times 27
\end{aligned}
$$

Therefore, $(-4, 27)$ is one special solution to the equation

$$155x + 23y = 1$$

Or, equivalently, $(-4 \times 9, 27 \times 9) = (-36, 243)$ is one special solution to *Equation 6.10*:

$$155x + 23y = 9$$

Because $\gcd(155, 23) = 1$, the general solution is given by:

$$
\begin{cases}
x &= -36 - 23 \cdot t \\
y &= 243 + 155 \cdot t
\end{cases}
\tag{6.11}
$$

where t is an integer.

Equation 6.11 is correct. However it is possible to simply it by reducing the values of these two constants.

Note that $\left[\frac{36}{23}\right] = 1$ and $\left[\frac{243}{155}\right] = 1$, where $[x]$ denotes the largest integer not exceeding x, *Equation 6.11* can be rewritten as:

$$\begin{cases} x & = -13 - 23 \cdot (t+1) \\ \\ y & = 88 + 155 \cdot (t+1) \end{cases}$$

Hence, its general solution can also be written as:

$$\begin{cases} x & = -13 - 23 \cdot t \\ \\ y & = 88 + 155 \cdot t \end{cases} \tag{6.12}$$

Done.

Simplifying *Equation 6.11* to 6.12 is not mandatory, but beneficial.

Example 6.3.2

Let positive integers a and b satisfy $\gcd(a, b) = 1$. Show that the number of non-negative solutions to the equation

$$ax + by = c$$

is either $\left[\dfrac{c}{ab}\right]$ or $\left[\dfrac{c}{ab}\right] + 1$.

This can be proved rigidly using algebra. Here, we just provide a geometric explanation and leave the complete proof to the readers.

Geometrically, the equation $ax + by = c$ represents a straight line as shown below. Therefore, the number of non-negative integer solutions equals the number of grid points on the line in quadrant *I*. A grid point is defined as a point whose x and y coordinates are both integers.

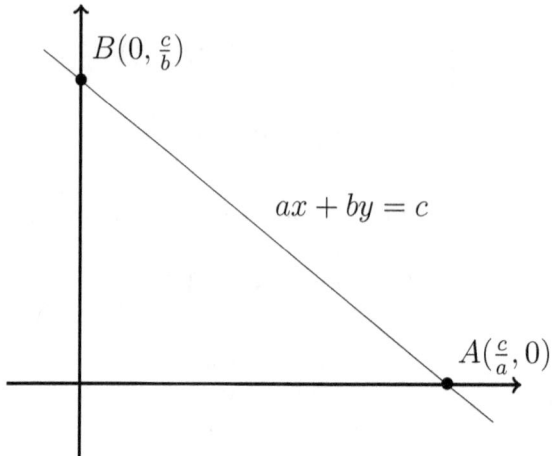

This line's X-intercept and Y-intercept are $\frac{c}{a}$ and $\frac{c}{b}$, respectively. It is clearly that the x coordinate of every point on segment AB satisfies $0 \le x \le [\frac{c}{a}]$. Within this range, there are either $[\frac{c}{a}]$ or $[\frac{c}{a}]$ $+ 1$ integers, depending on whether a is divisible by c, i.e. whether point A is a grid point, or equivalently, whether $a \mid c$ holds or not.

For every point whose X-coordinate is x, its corresponding Y-coordinate equals $\frac{c - a \cdot k}{b}$. To ensure both x and y are integers, b must be divisible by $(c - ak)$. Because a and b are relatively prime, there are only one integer satisfying this condition in every b numbers. Hence the total number of grid points whose coordinates are both integers is either $[\frac{c}{ab}]$ or $[\frac{c}{ab}] + 1$.

6.4   Multi-variable Equations

Conceptually, a variable represents one degree of freedom, or one moving part. An equation is one constraint that regulates the relationship among these variables. To solve an equation is to find such a system's balance point.

Generally speaking, for a system to get balanced, the number of moving parts and the number of constraints usually need to be equal. In *Example 5.3.1*, there are two variables with just one equation. As a result, its general solution contains one parameter which represents the missing link.

> The number of parameters usually equals the difference between the number of variables and the number of equations.

For example, if there are three variables to solve, but just one equation is given, two parameters are usually required.

The process to solve a multi-variable equation is to eliminate variables one by one, and meanwhile, to introduce parameters one by one accordingly. This is illustrated in the following example:

Example 6.4.1

Solve the following equation in integers:

$$189x + 180y + 175z = 15750$$

This equation contains three variables. Hence two parameters are required. In the solution below, we first eliminates z by introducing u as the 1st parameter. This is followed by eliminating y and introducing v as the 2nd parameter.

Solution

Let's first solve (i.e. eliminate) z by assuming x and y are known:

$$z = \frac{15750 - 198x - 180y}{175}$$

$$z = \frac{175 \times 90 - 175x - 14x - 175y - 5y}{175}$$

$$z = 90 - x - y + \frac{-14x - 5y}{175} \tag{6.13}$$

As x, y, z are all integers, $\frac{-14x-5y}{175}$ must be an integer too. Let it be u:

$$u = \frac{-14x - 5y}{175} \qquad (6.14)$$

Note that *Equation 6.14* does not contain z. Instead, it contains a parameter u which, conceptually, can be treated as a known quantity. Rearranging this new equation leads to:

$$14x + 5y + 175u = 0$$

Next, solving (i.e. eliminating) y with respect to x and u:

$$y = -35u - 3x + \frac{x}{5} \qquad (6.15)$$

We now find $\frac{x}{5}$ must be an integer. Let it be v: $v = \frac{x}{5}$.

This means $x = 5v$. Substituting x with $5v$ in *Equation 6.15* yields:

$$y = -35u - 14v$$

Finally by *Equation 6.13*, we have

$$z = 90 - x - y + u = 90 + 9v + 36u$$

Therefore the solution to the original equation is:

$$\begin{cases} x & = 5v \\ y & = -35u - 14v \\ z & = 90 + 36u + 9v \end{cases}$$

where u and v are two integer parameters.

Done.

It is possible to solve this equation by following an alternative variable elimination process, e.g. to eliminate x first, followed by y, and so on. This will lead to different but equivalent solutions.

6.5   Practice

Practice 1

Find the general solution to the equation

$$37x + 107y = 25 \qquad (6.16)$$

where both x and y are integers.

Practice 2

Find all positive integer solutions to the following equation:

$$7x + 19y = 213$$

Practice 3

Show that any triplet of (x, y, z) that is generated by the following parametric expressions can form a right triangle.

$$\begin{cases} x & = m^2 - n^2 \\ y & = 2mn \\ z & = m^2 + n^2 \end{cases}$$

where m and n are two positive integers, and $m > n$.

Practice 4

Mary has collected 14 beads of three different types, each of which type weights 12 grams, 8 grams, and 5 grams individually. If the total weight of her collection is exactly 100 grams, what are the quantities of each kind?

Practice 5

Write 118 as the sum of two positive integers, one of which is a multiple of 11, and the other is a multiple of 17.

Practice 6

Let a, b and n be three positive integers, and $\gcd(a, b) = 1$, show that:

(ii) if $n > ab - a - b$, the equation $ax + by = n$ is solvable in non-negative integers

(iiii) if $n = ab - a - b$, the above equation is not solvable in non-negative integers

Chapter 7

Pythagorean Triplets

7.1 Pythagorean Triplets Defined

A positive integer triple (x, y, z) that satisfies the following relationship

$$x^2 + y^2 = z^2$$

is called a Pythagorean triplet because of its relationship with the Pythagorean theorem.

When x, y, and z are relatively prime to each other, (x, y, z) is called a *primitive* Pythagorean triplet.

There are totally 16 primitive Pythagorean triplets with $z < 100$.

(3, 4, 5)	(5, 12, 13)	(8, 15, 17)	(7, 24, 25)
(20, 21, 29)	(12, 35, 37)	(9, 40, 41)	(28, 45, 53)
(11, 60, 61)	(16, 63, 65)	(33, 56, 65)	(48, 55, 73)
(13, 84, 85)	(36, 77, 85)	(39, 80, 89)	(65, 72, 97)

Example 7.1.1

Let (x, y, z) be a primitive Pythagorean triplet. Show that x and y are of opposite parity.

Proof

Because x are y are relatively prime, it is clear that x and y cannot be both even.

They cannot be both odd either. Otherwise, if both x and y are both odd, then $z^2 = x^2 + y^2$ implies z is even, and thus $4 \mid z^2$.

However, on the other hand, if both x and y are odd, it must hold that $x^2 \equiv y^2 \equiv 1 \pmod 4$. Therefore, $z^2 = x^2 + y^2 \equiv 2 \pmod 4$. This is a contradiction to the previous assertion.

$$QED$$

It follows from *Example 7.1.1* that:

Property 7.1.1 Primitive Pythagorean Triplet

If (x, y, z) is a primitive Pythagorean triplet, then:

- x and y are opposite parity, and

- z is odd

Property 7.1.1 is useful when parity analysis is performed on Pythagorean triplet related problems.

7.2 Pythagorean Triplet Formula

Obviously, if (x, y, z) is a primitive Pythagorean triplet, then by multiplying a positive integer scaling factor k, (kx, ky, kz) is a Pythagorean triplet too

On the other hand, if (x, y, z) is a Pythagorean triplet, dividing x, y, and z by their greatest common divisor must lead to a corresponding primitive Pythagorean triplet.

Therefore, it is sufficient to focus on finding a general solution to primitive Pythagorean triplets. If all the primitive Pythagorean triplets can be constructed, all the Pythagorean triplets can be obtained too by applying appropriate multipliers.

One of the most used Pythagorean triplet formulas is given by:

$$\begin{cases} x = m^2 - n^2 \\ y = 2mn \\ z = m^2 + n^2 \end{cases} \tag{7.1}$$

where m and n are positive integers, and satisfy the following conditions:

(i) $m > n$, and

(ii) m and n are relatively prime, and

(iii) m and n are of opposite parity

When *Equation 7.1* is employed to solve a problem, these three conditions are frequently referenced and utilized.

Example 7.2.1

Show that in order to construct a primitive Pythagorean triplet using *Equation 7.1*, both m and n must be of opposite parity.

Proof

By *Property 7.1.1*, x and y are of opposite parity in order to qualify as a primitive triplet. Because $y = 2mn$ is even, x must be odd. It follows that m and n must have opposite parity because $x = m^2 - n^2$.

QED

Equation 7.1 can generate all Pythagorean triplets, primitive or non-primitive, by setting appropriate values to m and n. If they satisfy conditions (i) to (iii) above, the resulting triplets are primitive. For example:

m	n	x	y	z
2	1	3	4	5
3	2	5	12	13
4	1	15	8	17
...	...	...	...	...

Otherwise, if $\gcd(m, n) \neq 1$ or they have same parity, the resulting triplets are not primitive:

m	n	x	y	z
3	1	8	6	10
4	2	12	16	20
...	...	...	...	...

7.3 More Examples

For convenience, let's introduce the following definition:

> ### Definition 7.3.1   Pythagorean Triangle
>
> A triangle is called a Pythagorean triangle if it is a right triangle and its three sides' lengths are all integers.

Now let's consider several examples that utilize the Pythagorean formula.

Example 7.3.1

Find all Pythagorean triangles whose areas and perimeters are equal in values.

Solution

The lengths of three sides can be written as:

$$\begin{cases} x = m^2 - n^2 \\ y = 2mn \\ z = m^2 + n^2 \end{cases}$$

where m and n are two positive integers.

If its area and perimeter equal in values, the following must hold:

$$\frac{1}{2}(m^2 - n^2)(2mn) = (m^2 - n^2) + 2mn + (m^2 + n^2)$$

It follows that:

$$
\begin{aligned}
(m^2 - n^2)(mn) &= 2m^2 + 2mn \\
(m+n)(m-n)mn &= 2m(m+n) \\
(m-n)n &= 2
\end{aligned}
$$

This is a basic indeterminate equation that can be solved using the factorization method.

$$\begin{cases} m - n = 1 \\ n = 2 \end{cases} \quad \text{or} \quad \begin{cases} m - n = 2 \\ n = 1 \end{cases}$$

We find $(m, n) = (3, 2)$ or $(3, 1)$.

Setting these values into the Pythagorean triplet formula produces two such triangles: 5-12-13 and 6-8-10.

Done.

Example 7.3.2

A Pythagorean triangle must:

(i) have least one side whose length is a multiple of 3, and

(ii) have least one side whose length is a multiple of 4, and

(iii) have least one side whose length is a multiple of 5

Please note that these sides which satisfy the above conditions are not necessarily distinct. For example, in a 5-12-13 triangle, 12 is a multiple of both 3 and 4.

Proving *Example 7.3.2* requires some number theory knowledge. *Appendix A* on *page 101* provides a quick tutorial for those readers who are not familiar with that topic.

Here we prove the 1st statement. The remaining statements will be left as practice.

Proof

By the Pythagorean triplet formula, the lengths of these three sides can be written as:

$$\begin{cases} x = m^2 - n^2 \\ y = 2mn \\ z = m^2 + n^2 \end{cases}$$

where m and n are two positive integers.

If either m or n is a multiple of 3, then $y = 2mn$ must be a multiple of 3.

Otherwise if neither m nor n is a multiple of 3, then $x = m^2 - n^2$ must be a multiple of 3. This is because $k^2 \equiv 1 \pmod 3$ holds for any integer k which is not divisible by 3.[1]

$$QED$$

7.4 Fermat's Last Theorem

The equation of Pythagorean triplet $x^2 + y^2 = z^2$ turns out to be a special case of a more general form:

$$x^n + y^n = z^n \tag{7.2}$$

where n is a positive integer.

When $n = 1$, *Equation 7.2* becomes $x + y = z$. This is a trivial equation.

When $n = 2$, it becomes $x^2 + y^2 = z^2$. Its general solution is given by the Pythagorean triplet formula *Equation 7.1*.

It is then natural to ask what if $n > 2$? How to solve equations such as $x^3 + y^3 = z^3$, $x^4 + y^4 = z^4$, and so on? This was one subject that the great mathematician Pierre de Fermat had studied. In 1637, he conjectured a theorem which is now known as Fermat's Last Theorem. This theorem was not proved until 1995 after 358 years!

---

[1]Please refer to *Example A.3.5* on *page 107*.

> ### Theorem 7.4.1 Fermat's Last Theorem
>
> Let n be a positive integer, equation
>
> $$x^n + y^n = z^n$$
>
> has no integer solution if $n > 2$.

A complete proof of Fermat's Last Theorem is beyond the scope of this book. Having that said, some special cases can be proved using elementary methods, e.g. $n = 4$. Proving $x^4 + y^4 = z^4$ unsolvable is included as a practice in *Chapter 8*.

7.5 Practice

Practice 1

In the primitive Pythagorean formula below, what are the properties of x, y, z, m and n?

$$\begin{cases} x &= m^2 - n^2 \\ y &= 2mn \\ z &= m^2 + n^2 \end{cases}$$

Practice 2

Let integers a, b and c be the lengths of a Pythagorean triangle's three sides, where $c > a, b$. Show that

$$\frac{1}{2}(c - a)(c - b)$$

must be a square number.

Practice 3

Show that a Pythagorean triangle must:

(i) have least one side whose length is a multiple of 3, and

(ii) have least one side whose length is a multiple of 4, and

(iii) have least one side whose length is a multiple of 5

Note that these sides may not be necessarily distinct. For example, in a 5-12-13 triangle, the side 12 is a multiple of both 3 and 4.

Practice 4

How many grid points whose x and y coordinates are both integers locate on the circle centered at $(199, 0)$ with a radius of 199?

Chapter 8

The Infinite Descent Method

Infinite descent is an important proof method, along with the mathematical induction and proof by contradiction. It was first practiced by the great mathematician Pierre de Fermat. Solving indeterminate equations is one of its applications.

8.1 The Principle

The principle of this method is based on the following fact:

Theorem 8.1.1 Principle of Infinite Descent

For any given positive integer n, it is impossible to construct an *infinitive* series of positive integers a_1, a_2, a_3, $\cdots$ such that

$$n > a_1 > a_2 > a_3 > \cdots > 0$$

Its validity is obvious. There are exactly $n - 1$ positive integers between 0 and n. Hence, it is not possible to squeeze infinitely many distinct positive integers in this range.

This principle can be used in cooperation with proof by contradiction to show that some indeterminate equations have no positive integer solution. When this approach is employed, the following steps can be used as a guideline:

1. First, assume $S_0 = (x_0, y_0, z_0, \cdots)$ is one positive integer solution

2. Next, construct a new *smaller* positive integer solution $S_1 = (x_1, y_1, z_1, \cdots)$ based on $S_0 = (x_0, y_0, z_0, \cdots)$

3. If this process of constructing S_1 from S_0 is repeatable infinitively,

$$S_0 \implies S_1 \implies S_2 \implies \cdots$$

then it is impossible. This is because by *Theorem 8.1.1*, it is impossible to construct an infinitively decreasing positive integer sequence.

To qualify for a *smaller* solution as described in step 2 above, it is sufficient to have just one variable to form a strictly decreasing sequence. For example, if

$$x_0 > x_1 > x_2 > \cdots > 0 \tag{8.1}$$

then the solutions

$$(x_0, \cdots), (x_1, \cdots), (x_2, \cdots), \cdots \tag{8.2}$$

qualifies as a decreasing series regardless of the trends of other variables. This is because if no sequence *8.1* can exist, no series solutions *8.2* can exist as a result.

That being said, however, in most cases, all the variables will form decreasing sequences.

8.2   Examples

The infinite descent method is often used together with divisibility analysis. This technique is illustrated in the following example.

Example 8.2.1

Solve this equation in positive integers: $x^3 + 2y^3 = 4z^3$.

Solution

If there exists such a positive integer solution (x, y, z), then x must be even. Let $x = 2x_1$:

$$(2x_1)^3 + 2y^3 = 4z^3$$
$$4x_1^3 + y^3 = 2z^3$$

This means y must be even too. Let $y = 2y_1$:

$$4x_1^3 + (2y_1)^3 = 2z^3$$
$$2x_1^3 + 4y_1^3 = z^3$$

This in turn shows z is also even. Let be $z = 2z_1$:

$$2x_1^3 + 4y_1^3 = (2z_1)^3$$
$$x_1^3 + 2y_1^3 = 4z_1^3$$

This last equation is in the same form of the original one. Hence, we conclude if (x, y, z) is a positive integer solution, x, y, and z must be all even, and $(x_1, y_1, z_1) = (\frac{x}{2}, \frac{y}{2}, \frac{z}{2})$ will be a solution too. It is clear that the process of $(x, y, z) \implies (x_1, y_1, z_1)$ is repeatable. Therefore an infinitive decreasing solution series can be constructed, which is impossible by *Theorem 8.1.1*. Thus, the equation is unsolvable in positive integers.

Done.

The infinite descent method is not only useful in solving indeterminate equations, but also applicable to tackle other types of problems. Here is a classical one.

Example 8.2.2

Show that $\sqrt{2}$ is an irrational number.

Proof

If $\sqrt{2}$ is a rational number, let $\sqrt{2} = \frac{p}{q}$, where p and q are two positive integers that make this equation hold, and minimize the value of $(p+q)$.

Squaring both sides yields $2 = \frac{p^2}{q^2}$ or $2q^2 = p^2$. Therefore p must be even.

Setting $p = 2p_1$ leads to $2q^2 = (2p_1)^2$ or $q^2 = 2p_1^2$. This means q must be even too. Let it be $q = 2q_1$.

Hence we have

$$\sqrt{2} = \frac{p}{q} \implies \sqrt{2} = \frac{p_1}{q_1}$$

Obviously $p_1 + q_1 = \frac{p}{2} + \frac{q}{2} < p + q$ which contradicts the minimality assumption of $(p+q)$

$$QED$$

Example 8.2.2 also shows a useful variation of the indefinite descent method by assuming the starting positive integer solution (p, q) minimizes the value of $(p+q)$. If another positive integer solution (p_1, q_1) with a smaller benchmark $(p_1 + q_1)$ can be constructed from (p, q), then a contradiction is discovered.

8.3 Practice

Practice 1

Show that if a positive integer n is not a square number, then $\sqrt{n}$ is an irrational number.

Practice 2

Show that the area of a Pythagorean triangle cannot be a square number.

Practice 3

Show that $x^4 + y^4 = z^2$ is not solvable in positive integers.

Practice 4

Show that when $n = 4$, the Fermat's equation

$$x^n + y^n = z^n$$

is not solvable in positive integers.

Practice 5

Prove that if positive integers a and b are such that $ab+1$ divides $a^2 + b^2$. then

$$\frac{a^2 + b^2}{ab + 1}$$

is a square number.

(1988 IMO)

Chapter 8: The Infinite Descent Method

Chapter 9

Pell's Equation

In this chapter, we investigate indeterminate equations in the following form:

$$x^2 - dy^2 = \pm 1 \tag{9.1}$$

where d is a positive integer.

Such an equations is called a Pell's equation.

9.1 Introduction

It turns out that it matters whether d is a square number in *Equation 9.1* or not. When d is, its solution is trivial. Otherwise, its solution is more interesting. Hence, our focus is to study those cases when d is not a square number.

To be more precise, d is often assumed to be *square-free* which means none of its divisors, except 1, is a square number. This is because when d is divisible by a square number, the equation can always be transformed to another equivalent one whose corresponding d does not contain that square number divisor. To see this, let's assume $d = d_1 k^2$ where both d_1 and k are positive integers which are not equal to 1. Then we have:

$$x^2 - dy^2 = \pm 1 \implies x^2 - d_1(ky)^2 = \pm 1$$

Clearly, the 2nd equation is also a Pell's equation whose solutions have a one-to-one relationship with the original equation.

9.2 Trivial Cases

When d is a square number, solutions are trivial.

Example 9.2.1

Let d be a square number. Show that the equation $x^2 - dy = -1$ has only two integer solutions $(\pm 1, 0)$.

Proof

Let $d = k^2$, we have $x^2 - k^2 y^2 = 1 \implies (x + ky)(x - ky) = 1$.

Therefore one of the following systems must hold:

$$\begin{cases} x + ky & = 1 \\ x - ky & = 1 \end{cases} \quad \text{or} \quad \begin{cases} x + ky & = -1 \\ x - ky & = -1 \end{cases}$$

Solving these two systems leads to $(x, y) = (\pm 1, 0)$.

$$QED$$

Example 9.2.2

Show that when d is a square number, the equation $x^2 - dy = -1$ is solvable if and only if $d = 1$. In this case, its solutions are $(x, y) = (0, \pm 1)$.

Proof

Let $d = k^2$ where k is a positive integer, we have

$$x^2 - k^2y^2 = -1 \implies (x + ky)(x - ky) = -1$$

Therefore one of the following must hold:

$$\begin{cases} x + ky &= 1 \\ x - ky &= -1 \end{cases} \quad \text{or} \quad \begin{cases} x + ky &= -1 \\ x - ky &= 1 \end{cases}$$

It follows that $x = 0$ and $ky = \pm 1$. The later can only hold when $k = \pm 1$ and $y = \pm 1$. Hence we conclude the equation is only solvable in integers when $d = k^2 = 1$. The corresponding solutions are $(x, y) = (0, \pm 1)$.

$$QED$$

These two examples clearly show that solutions to Pell's equations are trivial when d is a square number.

9.3   Solutions to Pell's Equation

If d is a square-free positive integer[1], then the equation:

- $x^2 - dy^2 = 1$ always has infinitely many solutions, but

- $x^2 - dy^2 = -1$ may or may not have solutions

Let's first prove that the equation $x^2 - dy^2 = -1$ may be unsolvable in integer.

---

[1] *Square-free* means that d is not divisible by any square number that is not 1. Please see *Section 9.1*.

Example 9.3.1

Show that there exist infinitely many positive integers d such that the equation

$$x^2 - dy^2 = -1$$

does not have any integer solution.

Proof

Consider $d \equiv -1 \pmod 4$, and take $\pmod 4$ on both sides of the equation:

$$
\begin{aligned}
x^2 - dy^2 &\equiv -1 && \pmod 4 \\
x^2 - (-1)y^2 &\equiv -1 && \pmod 4 \\
x^2 + y^2 &\equiv 3 && \pmod 4
\end{aligned}
$$

However $x^2 + y^2 \equiv 3 \pmod 4$ will never hold for any integers x and y based on the conclusion of *Example A.3.7* on *page 108*.

This means when $d \equiv -1 \pmod 4$, the equation is not solvable in integer. Because there exist infinitely many such positive integers, the original claim holds.

QED

The equation $x^2 - dy^2 = -1$ may be unsolvable. But when it has at least one solution, it will have unlimited number of solutions. This will be discussed in *Section 9.3.2* later.

9.3.1 Solve Equation $x^2 - dy^2 = 1$

Pell's equations in the form of $x^2 - dy^2 = 1$, where d is a square-free positive integer, are always solvable. Because both terms of x^2 and y^2 are squares, if (x, y) is a solution, so will $(-x, -y)$ be. Consequently, we only need to focus on positive integer solutions.

Let's first consider the following example:

Example 9.3.2

Find one positive integer solution to each of the following equations:

(i) $x^2 - 2y^2 = 1$

(ii) $x^2 - 3y^2 = 1$

(iii) $x^2 - 5y^2 = 1$

Solution

Using the guessing-and-checking method, we find

(i) $(3, 2)$ is one solution to $x^2 - 2y^2 = 1$

(ii) $(2, 1)$ is one solution to $x^2 - 3y^2 = 1$

(iii) $(9, 4)$ is one solution to $x^2 - 5y^2 = 1$

Done.

It can be shown that the equation $x^2 - dy^2 = 1$ not only is solvable, but also has unlimited number of solutions. This conclusion is formalized in *Theorem 9.3.1* below.

Theorem 9.3.1

There exist infinitely many pairs of positive integers (x, y) such that $x^2 - dy^2 = 1$, where d is a square-free positive integer.

Among these solutions, the pair (x, y) that minimizes the value of $(x + \sqrt{d} \cdot y)$ is called its *fundamental solution*.

It is obvious that the fundamental solution to a given equation is unique. For example $(x, y) = (3, 2)$ is the fundamental solution to the equation $x^2 - 2y^2 = 1$.

While some fundamental solutions are easy to obtain, such as those in *Example 9.3.2*, some are difficult to determine. For instance, the fundamental solution to $x^2 - 13y^2 = 1$ is $(649, 180)$ which is not obvious at all. As such, it is convenient to compile a table of fundamental solution with respective to every d. Thus one can look them up when needed. This is possible because the equation $x^2 - dy^2 = 1$ is uniquely determined by d. Thus, for each d, its corresponding fundamental solution is deterministic. *Appendix B* contains such a table for $d \leq 30$.

Fundamental solution plays a critical role in determining all the solutions to a Pell's equation. Let (x_1, y_1) be the fundamental solution to the equation $x^2 - dy^2 = 1$. Then any pair of positive integer (x_n, y_n) obtained by the following relationship is one solution too.

$$x_n + \sqrt{d} \cdot y_n = (x_1 + \sqrt{d} \cdot y_1)^n \qquad (9.2)$$

On the other hand, any solution to the equation must satisfy *Equation 9.2* where n is a positive integer.

In order to understand *Equation 9.2*, let's expand the right side of the equation, and then rearrange them into the following form:

$$x_n + \sqrt{d} \cdot y = \underbrace{(\cdots)}_{A} + \sqrt{d} \cdot \underbrace{(\cdots)}_{B} \qquad (9.3)$$

where both part A and part B are polynomials of x_1 and y_1.

Because $\sqrt{d}$ is a radical number and all the other terms are integers, x_n must equal to part A and y_n must equal to part B, respectively, in order to make *Equation 9.3* hold. This fact leads to the following general solution:

Theorem 9.3.2   General Solution to $x^2 - dy^2 = 1$

The general solution to a Pell's equation $x^2 - dy^2 = 1$ where d is a square-free positive integer is given by:

$$
\begin{cases}
x_n = \dfrac{1}{2}\left((x_1 + \sqrt{d}\; y_1)^n + (x_1 - \sqrt{d}\; y_1)^n\right) \\[4mm]
y_n = \dfrac{1}{2\sqrt{d}}\left((x_1 + \sqrt{d}\; y_1)^n - (x_1 - \sqrt{d}\; y_1)^n\right)
\end{cases}
\tag{9.4}
$$

where (x_1, y_1) is the *fundamental solution*

Though *Equation 9.4* contains terms in radical forms, resulting x_n and y_n clearly are both integers.

Example 9.3.3

Solve the following equation in positive integers: $x^2 - 2y^2 = 1$.

Solution

Its fundamental solution is $(3, 2)$. Therefore its general solution is given by:

$$
\begin{cases}
x_n = \dfrac{1}{2}\left((3 + 2\sqrt{2})^n + (3 - 2\sqrt{2})^n\right) \\[4mm]
y_n = \dfrac{1}{2\sqrt{2}}\left((3 + 2\sqrt{2})^n - (3 - 2\sqrt{2})^n\right)
\end{cases}
$$

We can obtain the next two solutions by setting $n = 2$ and 3, respectively:

$$\left\{ \begin{array}{ll} x_2 = \dfrac{1}{2}\left((3+2\sqrt{2})^2 + (3-2\sqrt{2})^2\right) & = 17 \\[3ex] y_2 = \dfrac{1}{2\sqrt{2}}\left((3+2\sqrt{2})^2 - (3-2\sqrt{2})^2\right) & = 12 \end{array} \right.$$

$$\left\{ \begin{array}{ll} x_3 = \dfrac{1}{2}\left((3+2\sqrt{2})^3 + (3-2\sqrt{2})^3\right) & = 99 \\[3ex] y_3 = \dfrac{1}{2\sqrt{2}}\left((3+2\sqrt{2})^3 - (3-2\sqrt{2})^3\right) & = 70 \end{array} \right.$$

Done.

Equation 9.4 can be used to generate any solution by setting n to an appropriate value. However, the calculation can be tedious when n is not trivial. It turns out that there exists a recursive relationship between consecutive solutions as shown below. This can come handy when computing consecutive solutions are required or mathematical induction is used.

Theorem 9.3.3   Solutions to $x^2 - dy^2 = 1$ (Recursion)

Let (x_n, y_n) be solutions to the equation $x^2 - dy^2 = 1$ where d is a square-free positive integer. Then the following relationship holds:

$$\left\{ \begin{array}{ll} x_n & = x_1 \cdot x_{n-1} + d \cdot y_1 \cdot y_{n-1} \\ y_n & = x_1 \cdot y_{n-1} + y_1 \cdot x_{n-1} \end{array} \right. \tag{9.5}$$

where $n = 2, 3, 4, \cdots$

By utilizing *Equation 9.5*, we can compute solutions of the equation $x^2 - 2y^2 = 1$ in the following way:

Its fundamental solution (i.e. $n = 1$) is:

$$\begin{cases} x_1 & = 3 \\ y_2 & = 2 \end{cases}$$

Hence the recursive relationship is:

$$\begin{cases} x_n & = 3 \cdot x_{n-1} + 4 \cdot y_{n-1} \\ y_n & = 3 \cdot y_{n-1} + 2 \cdot x_{n-1} \end{cases}$$

Therefore:

$$\begin{cases} x_2 & = 3 \times 3 + 4 \times 2 & = 17 \\ y_2 & = 3 \times 2 + 2 \times 3 & = 12 \end{cases}$$

and

$$\begin{cases} x_3 & = 3 \times 17 + 4 \times 12 & = 99 \\ y_3 & = 3 \times 12 + 2 \times 17 & = 70 \end{cases}$$

These agree with previous answers given in *Example 9.3.3*.

9.3.2 Solve Equation $x^2 - dy^2 = -1$

We have already shown that the equation $x^2 - dy^2 = -1$ may be unsolvable in *Example 9.3.1* on *page 76*. However, there exist many cases when such an equation is solvable. For example, it can be shown that when d is a prime number and $d \equiv 1 \pmod 4$, it is always solvable.

Let's consider the following example:

Example 9.3.4

Find one positive integer solution to $x^2 - 5y^2 = -1$.

Solution

Using the guessing and checking method, we find $(x, y) = (2, 1)$ is one solution.

Done.

Furthermore, when the equation $x^2 - dy^2 = -1$ has at least one solution, it will have infinitely many solutions.

Theorem 9.3.4   Solutions to $x^2 - dy^2 = -1$

If a Pell's equation $x^2 - dy^2 = -1$ (where d is a square-free positive integer) has at least one positive integer solution, then it has infinitely many positive integer solutions. In such cases, its general solution is given by:

$$\begin{cases} x_n = \dfrac{1}{2}((x_1 + \sqrt{d}\, y_1)^{2n-1} + (x_1 - \sqrt{d}\, y_1)^{2n-1}) \\[2mm] y_n = \dfrac{1}{2\sqrt{d}}((x_1 + \sqrt{d}\, y_1)^{2n-1} - (x_1 - \sqrt{d}\, y_1)^{2n-1}) \end{cases} \tag{9.6}$$

where (x_1, y_1) is its *fundamental solution*.

Accordingly, the following relationship holds:

$$x_n + \sqrt{d}\, y_n = (x_1 + \sqrt{d}\, y_1)^{2n-1}$$

And its recursive relationship is given by:

Theorem 9.3.5 Solutions to $x^2 - dy^2 = -1$ (Recursion)

Let (x_n, y_n) be solutions to the equation $x^2 - dy^2 = -1$ where d is a square-free positive integer. Then the following relationship holds:

$$\begin{cases} x_n = (x_1^2 + d \cdot y_1^2) \cdot x_{n-1} + 2 \cdot d \cdot x_1 \cdot y_1 \cdot y_{n-1} \\ \\ y_n = 2 \cdot x_1 \cdot y_1 \cdot x_{n-1} + (x_1^2 + d \cdot y_1^2) \cdot y_{n-1} \end{cases}$$

(9.7)

where $n = 2, 3, 4, \cdots$

9.4 Alternative Recurrence Relationship

The recurrence relationship given in *Equation 9.5* and *9.7* are cross referenced. This means that x_n depends on both x_{n-1} and y_{n-1}. So does y_n.

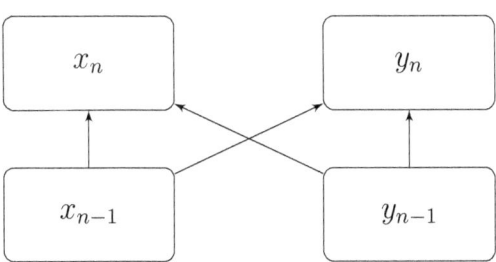

It is also possible to make x_n depend only on previous x_k's, and y_n only on previous y_k's.

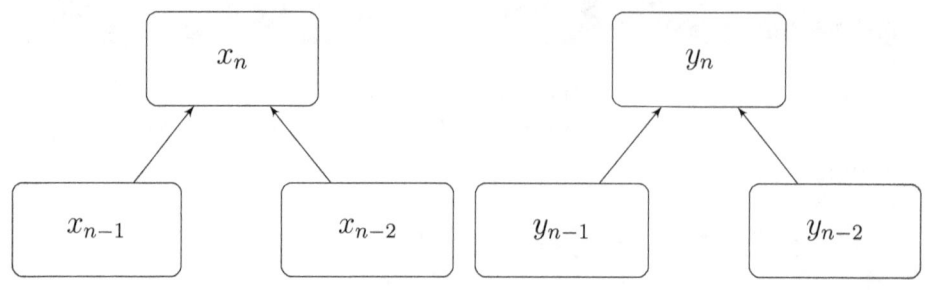

For $x^2 - dy^2 = 1$, the following relationship holds:

$$\begin{cases} x_n & = 2x_1 \cdot x_{n-1} - x_{n-2} \\ y_n & = 2x_1 \cdot y_{n-1} - y_{n-2} \end{cases} \qquad (9.8)$$

For $x^2 - dy^2 = 1$, the following relationship holds:

$$\begin{cases} x_n & = 2 \cdot (2 \cdot x_1^2 + 1) \cdot x_{n-1} - x_{n-2} \\ y_n & = 2 \cdot (2 \cdot x_1^2 + 1) \cdot y_{n-1} - y_{n-2} \end{cases} \qquad (9.9)$$

9.5 More Examples

Pell's equation has many interesting properties. Those relevant at middle and high school levels have been covered by now. This section will bring a few examples that utilize these properties.

Example 9.5.1

Show that there exist infinitely many triples of consecutive integers each of which is a sum of two squares.

For example: $8 = 2^2 + 2^2$, $9 = 3^2 + 0^2$, and $10 = 3^2 + 1^2$

Proof

Consider triples of $x^2 - 1$, x^2, and $x^2 + 1$. Because the equation

$$x^2 - 2y^2 = 1 \qquad (9.10)$$

has infinitely many positive integer solutions, there exist infinitely many triples which satisfy:

$$x^2 - 1 = y^2 + y^2 \qquad x^2 = x^2 + 0^2 \qquad x^2 + 1 = x^2 + 1^2$$

Therefore, the claim holds[2].

QED

Example 9.5.2

Show that if (x, y) is an integer solution to $x^2 - 2y^2 = 1$, then x must be odd and y must be even.

Proof

It is obvious that x is odd because $x^2 = 2y^2 + 1$. Let $x = 2n + 1$. Then we have:

$$(2n + 1)^2 - 2y^2 = 1$$
$$4n^2 + 4n + 1 - 2y^2 = 1$$
$$2n^2 + 2n - y^2 = 0$$
$$y^2 = 2n^2 + 2n$$

It is clearly now that y must be even.

QED

Example 9.5.3

Show that the equation $x^2 + y^3 = z^4$ has infinitely many integer solutions.

---

[2]The first relationship $x^2 - 1 = y^2 + y^2$ is simply a transformation of *Equation 9.10*. The other two relationships are trivial.

Proof

The identity: $1^3 + 2^3 + \cdots + n^3 = \left(\dfrac{n(n+1)}{2}\right)^2$ yields:

$$\left(\frac{(n-1)n}{2}\right)^2 + n^3 = \left(\frac{n(n+1)}{2}\right)^2 \tag{9.11}$$

If there exists infinitely many positive integers n and k, such that the equation

$$\frac{n(n+1)}{2} = k^2 \tag{9.12}$$

holds, then the given equation $x^2 + y^3 = z^4$ has infinitely many positive integer solutions:

$$(x, y, z) = \left(\frac{n(n-1)}{2}, n, k\right)$$

Rewrite *Equation 9.12* as

$$n^2 + n = 2k^2$$
$$4n^2 + 4n = 8k^2$$
$$4n^2 + 4n + 1 = 8k^2 + 1$$
$$(2n+1)^2 = 2 \times (4k^2) + 1$$
$$(2n+1)^2 - 2 \times (2k)^2 = 1 \tag{9.13}$$

Equation 9.13 is a Pell's equation in the form of

$$x^2 - 2y^2 = 1 \tag{9.14}$$

which has infinitely many solutions. By *Example 9.5.2*, x is even and y is odd in *9.14*. Therefore, any distinct solution to *9.14* can generate a distinct solution to *9.13*. In another word, *Equation 9.13* has infinitely many positive integer solutions. Therefore, the claim is true.

QED

Example 9.5.4

Find all triangles whose sides' lengths are consecutive integers and areas are also integers.

Solution

Let three sides' lengths be $z-1$, z, and $z+1$, respectively. Then by the Heron's formula, the triangle's area is given by:

$$S = \sqrt{\frac{3}{2}z \times \left(\frac{3}{2}z - (z-1)\right)\left(\frac{3}{2}z - z\right)\left(\frac{3}{2}z - (z+1)\right)}$$

$$= \frac{z}{4}\sqrt{3(z^2 - 4)} \tag{9.15}$$

If S is an integer, then $3(z^2 - 4)$ must be a square number. Let

$$3(z^2 - 4) = 3w^2$$

In addition, from *Equation 9.15*, it is clear that z must be even because, otherwise, both z and $\sqrt{3(z^2 - 4)}$ will be odd. This will make S a non-integer.

Letting $z = 2x$ leads to $4x^2 - 4 = 3w^2$. This means that w must be even too. Letting $w = 2y$ and simplifying yield:

$$x^2 - 3y^2 = 1$$

This is a Pell's equation. Its fundamental solution is

$$(x, y) = (2, 1)$$

and general solution is given by:

$$
\begin{cases}
x_n = \dfrac{(2 + \sqrt{3})^n + (2 - \sqrt{3})^n}{2} \\[4mm]
y_n = \dfrac{(2 + \sqrt{3})^n - (2 - \sqrt{3})^n}{2\sqrt{3}}
\end{cases}
\tag{9.16}
$$

As a result, there exist infinitely many such triangles. Three sides are $(2x_n - 1, 2x_n, , 2x_n + 1)$ where x_n is given by *Equation 9.16*, and area is $3 \cdot x_n \cdot y_n$.

The three smallest such triangles can be obtained by setting $n = 1, 2$, and 3, respectively:

$$(3, 4, 5), (13, 14, 15), (51, 52, 53)$$

Done.

9.6   Practice

Practice 1

Why is it usually assumed that d is not a square number when discussing the Pell's equation $x^2 - dy^2 = 1$?

Practice 2

Show that if integer pair (x, y) satisfies $x^2 - 2y^2 = -1$, then both x and y are odd.

Practice 3

Find all positive integer solutions to the equation:

$$x^2 - 5y^2 = 1$$

Practice 4

Show that if x and y are positive integer solutions to the equation $x^2 - 2y^2 = 1$, then $6 \mid xy$.

Practice 5

Solve in integers the equation $x^2 + y^2 - 1 = 4xy$.

Practice 6

Solve in *rational* numbers the equation $x^2 - dy^2 = 1$ where d is an integer.

Practice 7

Find all positive integers k, m such that $k < m$ and

$$1 + 2 + \cdots + k = (k+1) + (k+2) + \cdots + m$$

Practice 8

Let r be a positive real number, and $[r]$ be the largest integer not exceeding r. Prove that there exist unlimited number of positive integers, n, such that $[\sqrt{2}\, n]$ is a perfect square.

Chapter 10

The MOD Method

Modular arithmetic is an important topic in number theory. It studies and operates on remainders (*residue*) with respect to a given divisor (*modulo*). For those readers who are not familiar with modular arithmetic, *Appendix A* on *page 101* provides a quick tutorial.

10.1 When Consider Using MOD

The key to solve indeterminate equations is to reveal intrinsic relationships among and properties of variables. For example, in *Example 2.1.1* on *page 6*, the equation $x + xy + y = 8$ is solved based on the fact that both $(x + 1)$ and $(y + 1)$ are divisors of 9. The inequality method discussed in *Chapter 3* attempts to uncover such insight by determining certain variables' boundaries.

While these methods can tackle many types of indeterminate equations, they are usually not effective in handling exponential equations, such as

$$3^x + 4^y = 5^z \tag{10.1}$$

In such scenarios, the MOD method can be a better technique.

10.2 Solve Exponential Equations

The key to effectively employ the MOD method is to select appropriate modulo. Sometimes, this is not obvious, and thus requires careful observation and experience. That being said, there are some guidelines. The most important and practical one is to choose a modulo which can produce residues of 1, 0, or, -1 as many as possible. This is because these numbers are the easiest to deal with in exponential form.

Let's consider the following example:

Example 10.2.1

Show that if there exist positive integers x, y, and z satisfying the following equation

$$3^x + 4^y = 5^z \tag{10.2}$$

then both x and z must be even.

Proof

Taking (mod 4) on both sides of the equation leads to:
$$(-1)^x + 0 \equiv 1^z \pmod 4$$
Clearly, this relationship can only hold if x is even.

Next, taking (mod 3) on both sides yields:
$$0 + 1^y \equiv (-1)^z \pmod 3$$
Therefore, z must be even too.

$$QED$$

Though *Equation 10.2* is not completely solved, the MOD method reveals an important property of x and z. This information plays a vital role when we attempt to completely solve this equation later in this chapter.

Here is another example which also utilizes the power of the residue (-1).

Example 10.2.2

Find all ordered pairs of positive integers (x, y) that satisfy the equation:
$$17^x - 15^y = 2$$

Solution

One obvious solution is $(x, y) = (1, 1)$. We are going to show that this is the only solution.

If $y \geq 2$, then $15^y = 15^2 \times 15^{y-2}$ must be a multiple of 9. Taking (mod 9) on both sides of the equation:
$$(-1)^x - 0 \equiv 2 \pmod 9$$

Clearly, this relationship will never hold which means no solution exists for $y \geq 2$.

Done.

Listing obvious solutions first, and then investigating more complex situation, as illustrated in *Example 10.2.2*, is a useful technique when employing the MOD technique.

Example 10.2.3

Show that the equation $x^4 + 4 = 2y^4$ has no integer solution.

Proof

It is obvious that x^4 must be an even number. Therefore, x must be even too. It follows that $x^4 \equiv 0 \pmod{16}$. Hence the following holds for the left side of the equation:

$$x^4 + 4 \equiv 0 + 4 \equiv 4 \pmod{16}$$

However, for the right side:

- if y is even, then $2y^4 \equiv 0 \pmod{16}$

- if y is odd, then $2y^4 \equiv 2 \pmod{16}$

Either way, the two sides cannot equal. This means that no integer pairs (x, y) can satisfy the give equation.

$$QED$$

Both *Equation 10.2.2* and *10.2.3* have demonstrated that the MOD method is a powerful weapon to uncover intrinsic controversial from the remainder perspective, thus to show an equation is unsolvable.

However, using this method alone is usually insufficient to find all the solutions. In order to find all solutions in a constructive way, other methods are required. One frequently used technique is the factorization method, especially when exponents are shown to be even. This is illustrated in the following example:

Example 10.2.4

Find all positive integer solutions to the following equation

$$3^x + 4^y = 5^z$$

Solution

By employing the MOD method, we have already shown that both x and z must be even.[1] As such, let $x = 2k$, $z = 2p$, and note

---
[1]Please refer to *Example 10.2.1* on *page 92*.

$4^y = (2^y)^2$, the original equation becomes:

$$(3^k)^2 + (2^y)^2 = (5^p)^2$$

Therefore $(3^k, 2^y, 5^p)$ forms a Pythagorean triplet. Hence, there exist positive integers m and n such that [2]:

$$\begin{cases} 3^k &= m^2 - n^2 \\ 2^y &= 2mn \\ 5^p &= m^2 + n^2 \end{cases}$$

Because $2^y = 2mn$, both m and n must be some power of 2. Let $m = 2^t$ and $n = 2^s$ where t and s are non-negative integers satisfying $t + s = y - 1$. Note $m > n \implies t > s$.

It follows that:

$$\begin{cases} 3^k &= m^2 - n^2 &= 2^{2t} - 2^{2s} &= 2^{2s}(2^{2(t-s)} - 1) \\ 5^p &= m^2 + n^2 &= 2^{2t} + 2^{2s} &= 2^{2s}(2^{2(t-s)} + 1) \end{cases}$$

Because neither 3^k nor 5^p is divisible by 2, we conclude 2^{2s} must equal 1. This means $s = 0$, and $2^{2(t-s)} = 4$ or $t = 1$. It is followed by $k = p = 1$.

Hence, the given equation has only one positive integer solution: $x = y = z = 2$.

Done.

Casework is another frequently used facilitating technique when necessary. Let's consider the following example:

Example 10.2.5

Solve in positive integers the equation $2^x - 1 = y^z$.

---

[2]Note 3^k is an odd number. Therefore it cannot equal $2mn$

Solution

It is clear that the equation will become an identity in either of the following scenarios:

(i) If $x = y = 1$, then z can be any positive number.

(ii) If $z = 1$, then x can be any positive number because there always exists a y satisfying the relationship $y = 2^x - 1$

As a special case of the scenario (ii) above: if $x = 2$, then $y^z = 2^x - 1 = 3$ only holds if $y = 3$ and $z = 1$.

Consequently, if there exists any additional solution, x is at least 3. If follows that:

$$2^x - 1 \text{ is odd} \implies y^z \text{ is odd} \implies y \text{ is odd}$$

In addition,

$$x \geq 8 \implies y^z = 2^x - 1 \equiv -1 \pmod{8} \implies z \text{ is odd too}$$

This holds is because, given y is odd, if z is even, we must have $y^z \equiv 1 \pmod 8$. Please refer to *Example A.3.3* on *page 106* for a proof.

Hence, the original equation can be rewritten as:

$$2^x = y^z - 1 = (y - 1)\underbrace{(y^{z-1} + y^{z-2} - \cdots - y + 1)}_{A}$$

Part A has z terms, which is odd, and every term is odd. Therefore their sum must be odd. However, the left side 2^x is not divisible by any odd positive integer, except 1. Hence part A can only hold one term, 1, which means $z = 1$. The case $z = 1$ is already included in the two scenarios listed above. Hence, there is no additional solution possible.

Done.

10.3   Practice

Practice 1

Solve the following equation in integers: $x^2 + y^2 = 2015$.

Practice 2

Prove: $2x^2 - 5y^2 = 7$ has no integer solution.

Practice 3

Find all positive integer solutions to the equation $3^x - 2^y = 1$.

Practice 4

Solve this equation in positive integers: $|12^x - 5^y| = 7$.

Practice 5

Find all positive integer solutions to the equation

$$8^x + 15^y = 17^z$$

Practice 6

Solve in integers the equation $|3^x - 2^y| = 41$.

Practice 7

Find all positive integer solutions to $7^x - 3 \times 2^y = 1$.

Practice 8

Solve the following equation in positive integers:

$$2015x + 2000y + 3 = z^2$$

Appendices

Appendices

Appendix A

Quick Introduction to Modular Arithmetic

The purpose of this appendix is to give readers a quick introduction to modular arithmetic. Modular arithmetic is an important topic in number theory and has wide applications. It also forms an important foundation for solving some types of indeterminate equations, especially those in *Chapter 7* and *Chapter 10*.

A.1 Modular Arithmetic Defined

While not often being discussed in classrooms, remainder is a big and important topic in number theory. Modular arithmetic's purpose is to study remainders.

Let a, b, and r be some integers. For a positive integer m, if there exist integers k and l such that

$$\begin{cases} a & = m \cdot k + r \\ b & = m \cdot l + r \end{cases}$$

We say a and b are congruent modulo m, and write this relationship as:

$$a \equiv b \pmod{m}$$

Intuitively, this means a and b have the same remainder when being divided by m. Please note that the *remainder* here can be any integer, not necessarily satisfying $0 \leq r < m$. To avoid confusing with the usual definition used in classroom, it is often referred as *residue*.

Example A.1.1

$$20 \equiv 2 \equiv -1 \pmod{3}$$

It is worth noting that negative residues, especial -1, are frequently used in modular arithmetic to simplify calculation.

A.2   Modular Arithmetic Properties

Modular arithmetic shares many properties with the usual arithmetic.

Addition Subtraction and Multiplication
If $a \equiv b \pmod{m}$ and $c \equiv d \pmod{m}$, then
• $a \pm c \equiv b \pm d \pmod{m}$
• $a\,c \equiv b\,d \pmod{m}$

These two properties can be proved using the basic definition. Let's prove the 1st property here. Readers are encouraged to prove the 2nd one as an exercise .

Proof

As $a \equiv b \pmod{m}$, let $a = mk + r$ and $b = ml + r$, where k, l, and r are integers.

As $c \equiv d \pmod{m}$, let $c = mp + s$ and $d = mq + s$, where p, p, and s are integers.

Therefore $a \pm c = m(k \pm p) + (r \pm s)$, $b \pm d = m(l \pm q) + (r \pm s)$. This implies that when being divided by m, $(a \pm c)$ and $(b \pm d)$ shares the same residue, $(r \pm s)$, i.e.

$$a \pm c \equiv b \pm d \pmod{m}$$

QED

Because exponentiation can be treated as repeated multiplication, it is also possible to apply k^{th} power on both sides of a MOD equation.

$$a \equiv b \pmod{m} \implies \underbrace{a \cdots a}_{k} \equiv \underbrace{b \cdots b}_{k} \pmod{m}$$

Exponential

Let k be a positive integer. If $a \equiv b \pmod{m}$, then

$$a^k \equiv b^k \pmod{m}$$

Though simple and intuitive, these properties can be useful in solving some problems.

Example A.2.1

Find the remainder when 20^{2015} is divided by 7.

Solution

$$20^{2015} \equiv (-1)^{2015} \equiv -1 \equiv 6 \pmod{7}$$

Therefore the answer is 6.

Done.

Example A.2.1 also demonstrates the power of using (-1) in intermediate steps. Please note, though it is acceptable to use negative values in the intermediate steps, the final answer must be positive if the question asks for remainders.

> However, while addition, subtraction, and multiplication properties all hold, division property usually does not hold.

For example, while $16 \equiv 10 \pmod{6}$ and $2 \equiv 2 \pmod{6}$, we cannot conclude that $16 \div 2 \equiv 10 \div 2 \pmod{6}$. In order to apply the division property, we have to divide the modulo by the greatest common factor of the two numbers in question:

Division

If $a \equiv b \pmod{m}$ and $c \equiv d \pmod{m}$, then

$$\frac{a}{c} \equiv \frac{b}{d} \pmod{\frac{m}{gcd(a,b)}}$$

where $gcd(a,b)$ denotes the greatest common factor of a and b.

A.3   Some Useful Conclusions

Modular arithmetic is a powerful tool to derive many useful number theory conclusions in a concise way. Some of these conclusions are used in this book. They are listed below.

Example A.3.1

Let n be an odd number, show that $n^2 \equiv 1 \pmod 4$.

Proof

$$n \text{ is odd} \implies n \equiv \pm 1 \pmod 4 \implies n^2 \equiv 1 \pmod 4$$

QED

Alternative Approach

If n is an odd number, let it be $n = 2k+1$, where k is an integer.

It follows that $n^2 = 4k^2 + 4k + 1 = 4(k^2 + k) + 1$. This means that the remainder is 1 when n^2 is divided by 4.

Done.

Comparing the two solutions, it is easy to see that the modular arithmetic based solution is more concise.

As an implication of *Example A.3.1*, we have the following conclusion:

Example A.3.2

Show that the difference of two odd integers' squares must be a multiple of 4.

Proof

Let m and n be two odd integers, then by *Example A.3.1*:

$$m^2 \equiv n^2 \equiv 1 \pmod 4 \implies m^2 - n^2 \equiv 0 \pmod 4$$

$$QED$$

This can also be proved using odd-even parity analysis:

$$m^2 - n^2 = (m+n)(m-n)$$

If both m and n are odd integers, then both $(m+n)$ and $(m-n)$ are even. As a result, the product of two even integers must be a multiple of 4.

In fact, the conclusion presented in *Example A.3.1* can become stronger, as shown in the following example:

Example A.3.3

Let n be an odd number, show that $n^2 \equiv 1 \pmod 8$.

Solution

Because n is an odd number, one of the following must hold:

$$n \equiv \pm 1 \pmod 8 \quad \text{or} \quad n \equiv \pm 3 \pmod 8$$

In the former case, $n^2 \equiv 1 \pmod 8$. In the later case, $n^2 \equiv 9 \equiv 1 \pmod 8$.

Therefore $n^2 \equiv 1 \pmod 8$ always holds, regardlessly.

Done.

Example A.3.4

Let n be a positive integer. Show that if n is not divisible by 3, then $n^2 \equiv 1 \pmod 3$.

Proof

As n is not divisible by 3, it must hold that $n \equiv \pm 1 \pmod 3$. It follows that
$$n^2 \equiv 1 \pmod 3$$

QED

An extension of *Example A.3.4* is:

Example A.3.5

Let $m > n$ be two positive integers. If neither m nor n is a multiple of 3, then $m^2 - n^2$ must be a multiple of 3.

Proof

By the conclusion of *Example A.3.4*:
$$m^2 \equiv n^2 \equiv 1 \pmod 3 \implies m^2 - n^2 \equiv 0 \pmod 3$$

QED

Example A.3.6

Let $m > n$ be two positive integers. If neither m nor n is a multiple of 5, then either $m^2 + n^2$ or $m^2 - n^2$ is a multiple of 5.

Proof

First, let's prove if k is not divisible by 5, then either $k^2 \equiv 1$ (mod 5) or $k^2 \equiv 4$ (mod 5) holds. This statement is true because k must be in one of the following forms: $k \equiv \pm 1$ (mod 5) and $k \equiv \pm 2$ (mod 5). In the former case, $k^2 \equiv 1$ (mod 5), and in the later case, $k^2 \equiv 4$ (mod 5).

This means the remainder of m^2 and n^2 must be either 1 or 4, when being divided by 5. If their remainders are the same, then $m^2 - n^2$ must be a multiple of 5. Otherwise, if their remainders are different, then $m^2 + n^2$ must be a multiple of 5.

QED

Example A.3.7

Let n be a sum of two squares, show that $n \equiv 3$ (mod 4) will never hold.

Proof

For any positive integer k, it must hold $k^2 \equiv 0$ or 1 (mod 4).

Therefore if n is the sum of two squares, the remainder of n being divided by 4 can only be $0 + 0 = 0$, $0 + 1 = 1$, or $1 + 1 = 2$. Hence $n \equiv 3$ (mod 4) will never hold.

QED

Appendix B

Pell's Equation: Table of Fundamental Solutions

The following table lists the fundamental solutions to the Pell's Equation:

$$x^2 - dy^2 = 1$$

where positive integer $d \leq 30$ is not a square number.

d	x	y	d	x	y	d	x	y
1	-	-	11	10	3	21	55	12
2	3	2	12	7	2	22	197	42
3	2	1	13	649	180	23	24	5
4	-	-	14	15	4	24	5	1
5	9	4	15	4	1	25	-	-
6	5	2	16	-	-	26	51	10
7	8	3	17	33	8	27	26	5
8	3	1	18	17	4	28	127	24
9	-	-	19	170	39	29	9801	1820
10	19	6	20	9	2	30	11	2

Appendix C

Solutions

C.1　Introduction

Practice 1

What is an indeterminate equation?

An equation becomes indeterminate when there are more than one variable to solve. (Or when the number of variables exceeds the number of the given equations.)

Practice 2

What are the objectives when working on an indeterminate equation?

In order to claim an indeterminate equation is solved, one of the following must be achieved:

- Prove the given equation has no solution, or

- Enumerate all the solutions, or

- Find its general solution.

Practice 3

What do you think is the key to solve an indeterminate equation?

The key to solve an indeterminate equation is to uncover intrinsic relationship among and properties of variables. Almost all the solving techniques are to reveal such information.

For example, the key to solve the equation $x + xy + y = 8$ is to uncover the fact that both $(x+1)$ and $(y+1)$ are divisors of 9. This will be discussed later.

C.2 The Factorization Method

Solve this equation in integers $3(x + y) = xy + 8$.

Re-write the given equation as the following:
$$3x + 3y = xy + 8$$
$$3x + 3y - xy = 8$$
$$xy - 3x - 3y = -8$$
$$xy - 3x - 3y + 9 = 1$$
$$(x - 3)(y - 3) = 1$$

It follows that:
$$\begin{cases} x - 3 = \pm 1 \\ y - 3 = \pm 1 \end{cases}$$

Therefore, the solutions to this equation are $(4, 4)$ and $(2, 2)$.

A line $y = px$, where p is a non-zero integer, intersects line $y = x + 10$ at a grid point whose x and y coordinates are both integers. How many such lines $y = px$ exist?

The answer is equal to the count of the paired integers (p, x) such that $px = x + 10$, where $p \neq 0$.

This equation is equivalent to $(p - 1)x = 10$.

Because 10 has totally 8 divisors, including negative ones, the above equation has 8 solutions. However one solution $(p - 1, x) = (-1, -10)$ will result in $p = 0$ which should be excluded. Hence the final answer is 7.

Practice 3

How many ordered integer pairs (x, y) are there that can satisfy the following equation?

$$\frac{x+3}{x+1} - y = 0$$

The given equation is equivalent to

$$1 + \frac{2}{x+1} - y = 0 \qquad \text{(C.1)}$$

or

$$(x+1)(y-1) = 2$$

The number of integer solutions equals the number of divisors that 2 has, including negative divisors. Therefore the answer is 4.

An alternative is to ensure that $\frac{2}{1+x}$ is an integer because all the other terms are integers in *Equation C.1*. It follows that $(1 + x)$ must be a divisor of 2. Because 2 has 4 divisors (including negative ones), therefore the answer is 4.

Practice 4

A grid point is defined as a point whose x and y coordinates are both integers. How many grid points will the function plot

$$y = \frac{x + 12}{2x - 1}$$

pass?

In this case, y cannot be expressed as the sum of an integer and a fraction where x only appears in the denominator. However, the polynomial factorization method can still be employed.

Rewrite the equation as

$$y(2x - 1) = x + 12$$
$$2xy - x - y = 12$$
$$4xy - 2x - 2y = 24$$
$$(2x - 1)(2y - 1) = 25$$

Because 25 has 6 divisors including negative ones, therefore the answer is 6.

Practice 5

Solve the following equation in positive integers:

$$\frac{1}{x} + \frac{1}{y} = \frac{1}{5}$$

The equation is equivalent to

$$5x + 5y = xy$$
$$5x + 5y - xy + 25 = 25$$
$$(x - 5)(y - 5) = 25$$

Because both x and y are integers, $(x - 5)$ and $(y - 5)$ will be integers too. Hence, they must be paired divisors of 25:

$x - 5$	$y - 5$	x	y
1	25	6	30
5	5	10	10
25	1	30	6

Practice 6

For each positive integer n, let $s(n)$ denote the number of ordered positive integer pair (x, y) for which

$$\frac{1}{x} + \frac{1}{y} = \frac{1}{n}$$

Find all positive integers n for which $s(n) = 5$.

(Indian Mathematical Olympiad)

Let n's prime factorization be

$$n = p_1^{a_1} p_2^{a_2} \cdots p_k^{a_k}$$

where integer $a_i \geq 1$ holds for every $i = 1, 2, \cdots k$.

By *Theorem 2.2.1* on *page 10*, we have

$$s(n) = (2a_1 + 1)(2a_2 + 1) \cdots (2a_k + 1) = 5$$

Obviously, every term inside a parenthesis satisfies $(2a_i + 1) > 1$. However, because 5 is a prime number, there can only be one such term, i.e.

$$s(n) = 2a_1 + 1 = 5$$

or $a_1 = 2$. This means that the prime factorization of n is:

$$n = p_1^2$$

where p_1 is prime number. In another word, n must be a square of a prime.

Practice 7

Find all integer pairs (x, y) such that $x - y^4 = 4$, where x is a prime number.

The given equation can be rewritten as:

$$x = y^4 + 4$$
$$x = y^4 + 4y^2 + 4 - 4y^2$$
$$x = (y^2 + 2)^2 - (2y)^2$$
$$x = (y^2 + 2 + 2y)(y^2 + 2 - 2y)$$
$$x = ((y+1)^2 + 1)((y-1)^2 + 1)$$

Because x is a prime number, the following relationship must hold:

$$\begin{cases} (y+1)^2 + 1 &= x \\ (y-1)^2 + 1 &= 1 \end{cases} \quad \text{or} \quad \begin{cases} (y+1)^2 + 1 &= 1 \\ (y-1)^2 + 1 &= x \end{cases}$$

Solving these two system leads to two solutions: $(5, \pm 1)$.

Practice 8

Let p and q be two distinct prime numbers. Solve the following equation in positive integers:

$$\frac{p}{x} + \frac{q}{y} = 1$$

The given equation is equivalent to

$$(x - p)(y - q) = pq$$

Because p and q are two distinct prime numbers, pq has four divisors: 1, p, q, and pq. Hence, one of the following must hold:

$$\begin{cases} x - p &= 1 \\ y - q &= pq \end{cases} \begin{cases} x - p &= p \\ y - q &= q \end{cases} \begin{cases} x - p &= q \\ y - q &= p \end{cases} \begin{cases} x - p &= pq \\ y - q &= 1 \end{cases}$$

Consequently, there are four positive integer solutions:

$$\begin{cases} x &= 1 + p \\ y &= p(1 + q) \end{cases} \begin{cases} x &= 2p \\ y &= 2q \end{cases} \begin{cases} x &= p + q \\ y &= p + q \end{cases} \begin{cases} x &= p(1 + q) \\ y &= 1 + q \end{cases}$$

Practice 9

For any given positive integer $n > 2$, show that there exists a right triangle satisfying the following conditions:

- the lengths of all its three sides are integers, and

- one of these lengths equals n

We only need to show that $x^2 + n^2 = z^2$ is solvable in positive integers when $n > 2$. This equation is equivalent to

$$(z - x)(z + x) = n^2$$

If n is odd, setting $(z - x, z + x) = (1, n^2)$ yields a positive integer solution

$$(x, z) = (\frac{n^2 - 1}{2}, \frac{n^2 + 1}{2})$$

If n is even, setting $(z - x, z + x) = (2, \frac{n^2}{2})$ yields a positive integer solution

$$(x, z) = (\frac{n^2 - 4}{4}, \frac{n^2 + 4}{4})$$

Hence, the equation $x^2 + n^2 = z^2$ is always solvable in positive integers.

C.3   The Inequality Method

Practice 1

Solve the following equation in integers: $y^2 = (x+1)(x+2)$.

Because $(x+1)^2 < y^2 < (x+2)^2$, this equation does not have any integer solution by the squeeze reasoning.

Practice 2

Solve in positive integers the equation: $\frac{1}{x} + \frac{1}{y} + \frac{1}{z} = 1$.

Without loss of generality, let's assume $x \geq y \geq z \geq 0$. Hence:

$$\frac{1}{x} \leq \frac{1}{y} \leq \frac{1}{z} \implies \frac{1}{z} \geq \frac{1}{3} \times 1 \implies z \leq 3$$

Do casework:

- If $z = 1 \implies \frac{1}{x} + \frac{1}{y} = 0 \implies$ no solution.

- If $z = 2 \implies \frac{1}{x} + \frac{1}{y} = \frac{1}{2} \implies \frac{1}{y} \geq \frac{1}{4} \implies 2 \leq y \leq 4$

  - if $y = 2 \implies \frac{1}{x} = 0 \implies$ no solution
  - if $y = 3 \implies \frac{1}{x} = \frac{1}{6} \implies x = 6 \implies (6, 3, 2)$
  - if $y = 4 \implies \frac{1}{x} = \frac{1}{4} \implies x = 4 \implies (4, 4, 2)$

- If $z = 3 \implies \frac{1}{x} + \frac{1}{y} = \frac{2}{3} \implies \frac{1}{y} \geq \frac{1}{3} \implies 3 \leq y \leq 3$

  - if $y = 3 \implies \frac{1}{x} = \frac{1}{x} \implies (3, 3, 3)$

Therefore, all the solutions are distinct permutations of

$$(6, 3, 2), (4, 4, 2), \text{ and } (3, 3, 3)$$

Practice 3

Find all positive integer triples (x, y, z) such that

$$3(xy + yz + zx) = 4xyz$$

Because x, y, and z are all positive integers, we can divide $3xyz$ on both sides of the equation and get

$$\frac{1}{x} + \frac{1}{y} + \frac{1}{z} = \frac{4}{3}$$

Without loss of generality, let's assume $0 \le x \le y \le z$. Hence:

$$\frac{1}{x} < \frac{1}{x} + \frac{1}{y} + \frac{1}{z} \le \frac{3}{x} \implies \frac{1}{x} < \frac{4}{3} \le \frac{3}{x} \implies 1 \le x \le 2$$

This means x can only be 1 or 2.

- $x = 1 \implies \frac{1}{y} + \frac{1}{z} = \frac{1}{3} \implies (y, z) = (6, 6), (4, 12)$

- $x = 2 \implies \frac{1}{y} + \frac{1}{z} = \frac{5}{6} \implies (y, z) = (2, 3)$

Hence, the solutions are all the distinct permutations of

$$(1, 4, 12), (1, 6, 6), \text{ and } (2, 2, 3)$$

Practice 4

Solve the following equation in integers:

$$x + \frac{1}{y + \frac{1}{z}} = \frac{10}{7}$$

The given equation is equivalent to:

$$x + \frac{z}{1+yz} = \frac{10}{7} \tag{C.2}$$

We note that y and $1 + yz$ are relatively prime, therefore $\frac{y}{1+yz}$ must be a fraction in its simplest form. This implies

$$1 + yz = \pm 7 \implies |z| \leq 8$$

Replacing $1 + yz$ with ± 7 in *Equation C.2* yields:

$$x = \frac{10 \pm z}{7} \implies 7 \mid 10 \pm z$$

Therefore z can only be 3 or 4. As a result, there are only two solutions:

$$(x, y, z) = (1, 2, 3), (2, -2, 4)$$

Practice 5

Solve the following equation in positive integers

$$\left(1 + \frac{1}{x}\right)\left(1 + \frac{1}{y}\right)\left(1 + \frac{1}{z}\right) = 2$$

(UK Mathematical Olympiad)

By symmetry, let's assume $x \geq y \geq z > 0$. Then, it is obvious that

$$\left(1 + \frac{1}{z}\right)^3 \geq \left(1 + \frac{1}{x}\right)\left(1 + \frac{1}{y}\right)\left(1 + \frac{1}{z}\right) = 2$$

This means $0 < z \leq 3$. Do casework:

(i) $z = 1 \implies (1 + \frac{1}{x})(1 + \frac{1}{y}) = 1 \implies$ no solution

(ii) $z = 2 \implies (1 + \frac{1}{x})(1 + \frac{1}{y}) = \frac{4}{3}$

$$\implies (x, y, z) = (15, 4, 2), (9, 5, 2), (7, 6, 2)$$

(iii) $z = 3 \implies (1 + \frac{1}{x})(1 + \frac{1}{y}) = \dfrac{3}{2}$

$$\implies (x, y, z) = (8, 3, 3), (5, 4, 3)$$

Therefore, we find that the solutions are all the distinct permutations of the following set:

$$(7, 6, 2), (9, 5, 2), (15, 4, 2), (8, 3, 3) \text{ and } (5, 4, 3)$$

Practice 6

Let integers a, b, and c satisfy $a - 2b = 4$ and $ab + c^2 - 1 = 0$. Find the value of $a + b + c$.

Canceling a from the two give equations leads to:

$$2b^2 + 4b + c^2 - 1 = 0$$

This relationship can turn into the sum of two perfect squares:

$$2(b + 1)^2 + c^2 = 3$$

Therefore,

- $b + 1 = \pm 1 \implies b = -2, 0$

- $c = \pm 1$

- $a = 2b + 4 \implies a = 0, 4$

Accordingly $a + b + c \in \{-1, -3, 3, 5\}$.

Practice 7

Find all positive integers n and k_i ($1 \leq i \leq n$) such that

$$k_1 + k_2 + \cdots + k_n = 5n - 4$$

and

$$\frac{1}{k_1} + \frac{1}{k_2} + \cdots + \frac{1}{k_n} = 1$$

(Putnam Mathematical Competition)

By the Cauchy-Schwarz inequality, we have

$$(k_1 + k_2 + \cdots + k_n)\left(\frac{1}{k_1} + \frac{1}{k_2} + \cdots + \frac{1}{k_n}\right) \geq n^2$$

Therefore it must hold that $(5n - 4) \geq n^2$, which means $n \leq 4$.

Do casework:

(i) when $n = 1$: we must have $k_1 = 1$

(ii) when $n = 2$:

This leads to $k_1 + k_2 = 6$ and $\frac{1}{k_1} + \frac{1}{k_2} = 1$. This system does not have any integer solution.

(iii) when $n = 3$:

This leads to $k_1 + k_2 + k_3 = 11$ and $\frac{1}{k_1} + \frac{1}{k_2} + \frac{1}{k_3} = 1$.

Solutions to this system are all the distinct permutations of $(2, 3, 6)$.

(iv) when $n = 4$:

This will make the equality $(5n - 4) = n^2$ hold. Therefore:

$$k_1 = k_2 = k_3 = k_4 \implies k_1 = k_2 = k_3 = k_4 = 4$$

Therefore, in conclusion, all the solutions are:

- $n = 1, k_1 = 1$

- $n = 3$, (k_1, k_2, k_3) are distinct permutations of $(2, 3, 6)$

- $n = 4, k_1 = k_2 = k_3 = k_4 = 4$

Practice 8

Solve the following equation in positive integers:

$$xy + yz + zx - xyz = 2$$

When there are terms such as xy, yz, zx and xyz appearing together, a common technique is to divide all of them by xyz. However in this case, there is a constant term, 2, let's use substitution to eliminate it first.

Let $u = x - 1$, $v = y - 1$, and $w = z - 1$, then we have

$$u + v + w = uvw \qquad \text{(C.3)}$$

where u, v, and w are non-negative integers because x, y, and z are positive integers.

If $uvw = 0$, i.e. at least one of them equals 0, then the other two must be opposite numbers. By symmetry, let's assume $u = 0$. This leads to the following solutions:

$$(u, v, w) = (0, k, -k)$$

where k is any integer. Because both v and w are non-negative, k must equal 0. Accordingly,

$$(x, y, z) = (1, 1, 1)$$

Otherwise, if $uvw \neq 0$, dividing uvw on both sides of *Equation C.3* yields:

$$\frac{1}{uv} + \frac{1}{vw} + \frac{1}{wu} = 1 \qquad \text{(C.4)}$$

Letting $r = uv$, $s = vw$, and $t = wu$ leads to the following standard form:

$$\frac{1}{r} + \frac{1}{s} + \frac{1}{t} = 1 \qquad (C.5)$$

Clearly, all of r, s, and t are positive integers.

By the *Practice C.3* on *page 120*, solutions to *Equation C.5* are all the distinct permutations of:

$$(r, s, t) = (6, 3, 2), (4, 4, 2), \text{ and } (3, 3, 3)$$

Because $rst = (uvw)^2$, only the product of $6, 3$, and 2 is square. Accordingly, (u, v, w) are all the distinct permutation of:

$$(u, v, w) = (1, 2, 3)$$

It follows that (x, y, z) are all the distinct permutation of

$$(x, y, z) = (2, 3, 4)$$

In conclusion, solutions are all the distinct permutations of $(2, 3, 4)$.

Practice 9

If integers x, y, and z are all greater than 2. Solve the following equation:

$$\frac{1}{x} + \frac{1}{y} - \frac{1}{z} = \frac{1}{2}$$

The given equation is equivalent to $\frac{1}{x}+\frac{1}{y}=\frac{1}{2}+\frac{1}{z}$. Therefore:

$$\frac{1}{x}+\frac{1}{y}>\frac{1}{2}$$
$$\frac{1}{x}>\frac{1}{2}-\frac{1}{y}$$
$$\frac{1}{x}>\frac{1}{2}-\frac{1}{3}$$
$$\frac{1}{x}>\frac{1}{6}$$
$$x<6$$

By the argument of symmetry, we can assert $y<6$ as well. Therefore, we only need to work on the following pairs of

$$(x,y)=(3,3),(3,4),(3,5),(4,4),(4,5),\text{ and }(5,5)$$

Among these choices, only three of them result in z being an integer:
$$(x,y,z)=(3,3,6),(3,4,12),\text{ and }(3,5,30)$$

Because x and y are symmetrical, and thus can be switched, all the solutions are:

$$(3,3,6),(3,4,12),(4,3,12),(3,5,30),\text{ and }(5,3,30)$$

Practice 10

Find all integers x that can satisfy the equation:

$$\frac{1}{x}+\frac{1}{x+1}+\frac{1}{x+2}=\frac{13}{12}$$

When $x<-2$, the left side will be negative, therefore no solution is possible. It is obvious that -2, -1, or 0 cannot be a solution as otherwise it will make one of the denominator 0.

When $x>0$, the left side is a strictly monotonically decreasing function. As such, we can start by finding an x which makes the left side approximately equal 0. If this equation is solvable, its solution must be near x.

Approximating the left side using $\frac{1}{x+1} + \frac{1}{x+1} + \frac{1}{x+1}$ leads to:

$$\frac{3}{x+1} = \frac{13}{12}$$

An approximate integer solution is $x = 2$. Setting $x = 2$ to the original equation finds it is the solution. Hence, this equation has a unique solution $x = 2$.

When a strictly monotonically increasing (or decreasing) function equals to a constant, there is at most one solution.

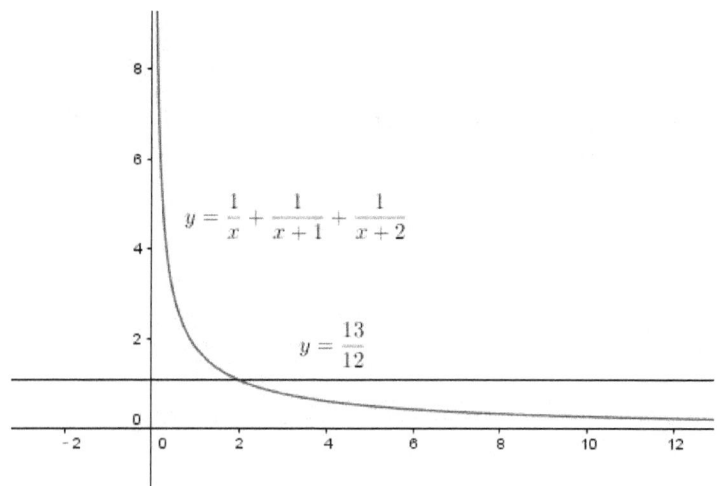

C.4   The Quadratic Method

Practice 1

Comparing with other techniques such as the factorization method, what are benefits of the quadratic method?

When the quadratic method can be used, its approach usually is systematic and requires less individualization. One will always start with organizing the equation as a quadratic one, then setting its discriminant as either non-negative or a square. By comparison, polynomial factorization often requires individualized maneuvers.

Practice 2

Solve this equation in integers: $y^2 - x^2 - 3x = 5$.

The given equation can be rewritten as:

$$x^2 + 3x + (5 - y^2) = 0$$

Hence
$$x = \frac{-3 \pm \sqrt{3^2 - 4 \times (5 - y^2)}}{2}$$

In order for x to be an integer, its discriminant

$$\Delta = 3^2 - 4 \times (5 - y^2) = 4y^2 - 11$$

must be a square of an odd number. Let it be

$$4y^2 - 11 = (2k + 1)^2$$

where k is a non-negative integer. It follows that

$$(2y + 2k + 1)(2y - 2k - 1) = 11$$

It is obvious that $2y + 2k + 1 > 2y - 2k - 1$ because k is non-negative. Therefore one of the following must hold:

$$\begin{cases} 2y + 2k + 1 = 11 \\ 2y - 2k - 1 = 1 \end{cases} \quad \text{or} \quad \begin{cases} 2y + 2k + 1 = -1 \\ 2y - 2k - 1 = -11 \end{cases}$$

Solving these two systems leads to $y = \pm 3$. Accordingly, the solutions to the original equations are:

$$(x, y) = (1, \pm 3) \text{ and } (-4, \pm 3)$$

Practice 3

Find all integer solutions to the following equation:

$$x^2 + 4xy + 5y^2 + 2x + 4y - 7 = 0$$

Treating the given equation as a quadratic equation of x. If this equation is solvable in integers, its discriminate must be non-negative:

$$\begin{aligned} \Delta &= (4y + 2)^2 - 4 \times (5y^2 + 4y - 7) \\ &= 16y^2 + 16y + 4 - 20y^2 - 16y + 28 \\ &= -4y^2 + 32 \\ &= 4 \times (8 - y^2) \\ &\geq 0 \end{aligned}$$

Therefore, $y^2 \leq 8 \implies y = (\pm 2, \pm 1, 0)$. Setting these values to the quadratic formula:

$$x = \frac{-(4y + 2) \pm \sqrt{\Delta}}{2} = -(2y + 1) \pm \sqrt{8 - y^2}$$

yields the following solutions:

$$(x, y) = (-3, 2), (-7, 2), (5, -2) \text{ and } (1, -2)$$

Practice 4

Solve the following equation in integers:

$$x^2 + xy + y^2 = 1$$

Rewrite the given equation with respect with x:

$$x^2 + xy + (y^2 - 1) = 0$$

In order for it to be solvable in integers, its discriminant must be non-negative:

$$\Delta = y^2 - 4 \times (y^2 - 1) = -3y^2 + 4 \geq 0$$

Hence, $y = 0, \pm 1$. Setting these values to the original equation yields the following solutions:

$$(x, y) = (\pm 1, 0), (0, \pm 1), (-1, 1), \text{ and } (1, -1)$$

Practice 5

Solve this equation in integers: $x^3 + y^3 = (x + y)^2$.

Obviously, when $x + y = 0$, or $(x, y) = (k, -k)$ where k is any integer, the equation holds.

Otherwise, dividing $(x + y)$ on both sides yields:

$$x^2 - xy + y^2 = x + y \tag{C.6}$$
$$x^2 - (y + 1)x + (y^2 - y) = 0 \tag{C.7}$$

In order to make both x and y integers, the discriminant must be non-negative:

$$\Delta = (y + 1)^2 - 4(y^2 - y) = -3y^2 + 6y + 1 \geq 0$$

or

$$1 - \frac{2\sqrt{3}}{3} \leq y < 1 + \frac{2\sqrt{3}}{3} \implies y = 0, 1, 2$$

Setting these values yield the following additional solutions:

$$(x, y) = (0, 1), (1, 0), (1, 2), (2, 1), \text{ and } (2, 2)$$

Note: this problem can also be solved using the sum of squares method. *Equation C.6* can be rewritten as:

$$(x - y)^2 + (x - 1)^2 + (y - 1)^2 = 2$$

Practice 6

Use the quadratic method to find all the integer pairs (x, y) that satisfy the equation:

$$(xy - 4)^2 = x^2 + y^2$$

This problem is the same as *Example 2.3.1* on *page 11*. It can also be solved using the quadratic method.

Rearranging the equation as a quadratic one with respect to x:

$$(y^2 - 1)x^2 - 8yx + (16 - y^2) = 0 \qquad \text{(C.8)}$$

For *C.8* to be solvable in integers, its discriminant must be a square:

$$\Delta = (-8y)^2 - 4(y^2 - 1)(16 - y^2)$$
$$= 4y^4 - 4y^2 + 64$$

Let it be k^2 where k is a non-negative integer:

$$4y^4 - 4y^2 + 64 = k^2$$
$$(2y^2 - 1)^4 + 63 = k^2$$
$$(k + (2y^2 - 1))(k - (2y^2 - 1)) = 3$$

Hence, one of the following must hold:

$$\begin{cases} k + (2y^2 - 1) & = \pm 63 \\ k - (2y^2 - 1) & = \pm 1 \end{cases} \quad \text{or} \quad \begin{cases} k + (2y^2 - 1) & = \pm 9 \\ k - (2y^2 - 1) & = \pm 7 \end{cases}$$

Solving this system yields the following integer solutions $y = \pm 4$ or 0. Setting these values to the original equation yields $x = \pm 4$. Therefore all the solutions are

$$(x, y) = (\pm 4, 0) \text{ or } (0, \pm 4)$$

C.5   The Euclidean Method

Practice 1

What is the necessary condition for the equation

$$ax + by = 1$$

to be solvable in integers, where both a and b are integers?

For this equation to be solvable in integers, a and b must be relatively prime.

Practice 2

What do you think is the necessary condition for the equation

$$ax + by = c$$

to be solvable in integers? Here, a, b and c are all integers, and c may or may not equal 1.

For this equation to be solvable in integers, c must be a multiple of the greatest common divisor of a and b, or $\gcd(a,b) \mid c$.

This conclusion will be discussed in the next chapter.

Practice 3

Create an equation in the form of $ax + by = 1$ yourself, and try to solve it.

Answer varies.

Practice 4

Show that for any given positive integer n, the fraction

$$\frac{7n+1}{14n+3}$$

must be in its simplest form.

This is equivalent to show that $(7n+1)$ and $(14n+3)$ are relatively prime which can be done by using the Bezout's method. In order to find two integers x and y satisfying the following equation

$$(7n+1)x + (14n+3)y = 1 \qquad\qquad \text{(C.9)}$$

rewrite it to an expression with respect to n:

$$(7x+14y)n + (x+3y) = 1 \qquad\qquad \text{(C.10)}$$

For *Equation C.10* to be an identity regardless of the value of n, the following must hold:

$$\begin{cases} 7x + 14y &= 0 \\ x + 3y &= 1 \end{cases}$$

Solving the above system yields $(x,y) = (-2,1)$. Therefore, there exist two integers $x = -2$ and $y = 1$ such that:

$$-2 \times (7n+1) + 1 \times (14n+3) = 1$$

or, *Equation C.9* is solvable in integers. Therefore, $(7n+1)$ and $(14n+3)$ are relatively prime.

C.6 General Solution

Find the general solution to the equation

$$37x + 107y = 25 \qquad (C.11)$$

where both x and y are integers.

The first step is to find one special solution to the equation

$$37x + 107y = 1 \qquad (C.12)$$

It can be done by guessing-and-checking, the Euclidean method or the modular equation method. Here we employ the Euclidean method.

$$107 = 37 \times 2 + 33$$
$$37 = 33 \times 1 + 4$$
$$33 = 4 \times 8 + 1$$

Hence:

$$1 = 33 - 4 \times 8$$
$$= 33 - (37 - 33 \times 1) \times 8$$
$$= 37 \times (-8) + 33 \times 9$$
$$= 37 \times (-8) + (107 - 37 \times 2) \times 9$$
$$= 37 \times (-26) + 107 \times 9$$

This means that $(-26, 9)$ is a solution to *Equation C.12*. It follows that $(-26 \times 25, 9 \times 25) = (-650, 225)$ is one solution to *Equation C.11*.

As $\gcd(37, 107) = 1$, the general solution is given by:

$$\begin{cases} x &= -650 - 107t \\ y &= 225 + 37t \end{cases}$$

where t is an integer.

Optionally, it is possible to reduce the constants in the general solution by noting:

$$650 = 107 \times 6 + 8 \text{ and } 225 = 37 \times 6 + 3$$

Therefore the general solution can be rewritten as:

$$\begin{cases} x &=& -8 - 107(t+6) \\ y &=& 3 + 37(t+6) \end{cases}$$

By substitution, we find a simpler general solution as the following:

$$\begin{cases} x &=& -8 - 107u \\ y &=& 3 + 37u \end{cases}$$

where u is an integer.

Practice 2

Find all positive integer solutions to the following equation:

$$7x + 19y = 213$$

Because $\gcd(7, 19) = 1$, the give equation is solvable in integers.

The 1st step is to find a special solution to $7x + 19y = 1$. Using the Euclidean method, we have:

$$19 = 7 \times 2 + 5$$
$$7 = 5 \times 1 + 2$$
$$5 = 2 \times 2 + 1$$

and

$$1 = 5 - 2 \times 2$$
$$= 5 - 2 \times (7 - 5 \times 1)$$
$$= 5 \times 3 - 7 \times 2$$
$$= (19 - 7 \times 2) \times 3 - 7 \times 2$$
$$= 7 \times (-8) + 19 \times 3$$

Hence on solution to $7x + 19y = 1$ is $(x, y) = (-8, 3)$. It follows that one solution to $7x + 19y = 213$ is

$$(x, y) = (-8 \times 213, 3 \times 213) = (-1704, 639)$$

Therefore the general solution to the equation $7x + 19y = 213$ is given by:

$$\begin{cases} x & = -1704 - 19t \\ y & = 639 + 7t \end{cases}$$

In order to reduce the constants, we note

$$1704 = 19 \times 89 + 13 \text{ and } 639 = 7 \times 91 + 2$$

It follows that

$$\begin{cases} x & = -1704 - 19(t - 90) & = 6 - 19(t - 90) \\ y & = 639 + 7(t - 90) & = 9 + 7(t - 90) \end{cases}$$

By substitution, the general solution can be written as

$$\begin{cases} x & = 6 - 19u \\ y & = 9 + 7u \end{cases}$$

where u is an integer.

Now we must ensure that both x and y are positive. Solving the following system:

$$\begin{cases} x & = 6 - 19u > 0 \\ y & = 9 + 7u > 0 \end{cases}$$

leads to $\dfrac{6}{19} > u > -\dfrac{9}{7}$.

There are only two integers in this range $u = 0, -1$. The corresponding solutions are:

$$(x, y) = (6, 9) \text{ and } (25, 2)$$

Practice 3

Show that any triplet of (x, y, z) that is generated by the following parametric expressions can form a right triangle.

$$\begin{cases} x &= m^2 - n^2 \\ y &= 2mn \\ z &= m^2 + n^2 \end{cases}$$

where m and n are two positive integers, and $m > n$.

It is sufficient to show that x, y, and z are three positive integers satisfying the Pythagorean theorem, i.e. $x^2 + y^2 = z^2$.

$$\begin{aligned} x^2 + y^2 &= (m^2 - n^2)^2 + (2mn)^2 \\ &= m^4 - 2m^2n^2 + n^4 + 4m^2n^2 \\ &= m^4 + 2m^2n^2 + n^4 \\ &= (m^2 + n^2)^2 \\ &= z^2 \end{aligned}$$

Because $m > n$ are two positive integers, it is clear that x, y, and z are all positive. Hence the claim holds.

Practice 4

Mary has collected 14 beads of three different types, each of which type weights 12 grams, 8 grams, and 5 grams individually. If the total weight of her collection is exactly 100 grams, what are the quantities of each kind?

Let x, y, and z be the quantities of each kind, respectively. Then

$$\begin{cases} 12x + 8y + 5z &= 100 \\ x + y + z &= 14 \end{cases}$$

Canceling z yields:

$$12x + 8y - 5(14 - x - y) = 100$$
$$7x + 3y = 30$$
$$y = 10 - \frac{7}{3}x$$

Therefore x must be a multiple of 3. In addition, both x and y must be positive integers. This results in $1 \leq x \leq 4$. It is obvious that $x = 3$ is the only candidate that satisfies both conditions.

Setting $x = 3$ leads to $(x, y, z) = (3, 3, 8)$.

Practice 5

Write 118 as the sum of two positive integers, one of which is a multiple of 11, and the other is a multiple of 17.

Let $11x$ and $17y$, where both x and y are integers, be the two integers that satisfy the condition. Then we have:

$$11x + 17y = 118$$

This is a standard equation in the form of $ax + by = c$. Let's solve it using the modular equation here.

$$11x + 17y = 118$$
$$x = \frac{118 - 17y}{11}$$
$$x = 10 - y + \frac{8 - 6y}{11}$$

Because both x and y are integer, $\frac{8-6y}{11}$ must be an integer, i.e.:

$$8 - 6y \equiv 0 \quad (\text{mod } 11)$$

Testing $y = 0, 1, \cdots, 10$ finds that $y = 5$ is one solution. Accordingly, $x = 3$. Therefore the general solution is:

$$\begin{cases} x & = 3 - 17t \\ y & = 9 + 11t \end{cases}$$

where t is an integer.

In order to make both x and y positive, t can only take 0. Thus, we conclude, the pair of positive integers that satisfies the requirement is 33 and 85.

Practice 6

Let a, b and n be three positive integers, and $\gcd(a, b) = 1$, show that:

(ii) if $n > ab - a - b$, the equation $ax + by = n$ is solvable in non-negative integers

(iiii) if $n = ab - a - b$, the above equation is not solvable in non-negative integers

Proof of statement (i)

Because $\gcd(a, b) = 1$, the equation $ax + by = n$ must be solvable in integers, and its general solution is given by:

$$
\begin{cases}
x &= x_0 - b \cdot t \\
y &= y_0 + a \cdot t
\end{cases}
$$

where (x_0, y_0) is one special solution.

Let's find a solution that satisfies $0 \leq x \leq b - 1$. This can be achieved by setting

$$t = \left[\frac{x_0}{b} \right]$$

In such a case,

$$x_0 - bt = x_0 - b \cdot \left[\frac{x_0}{b} \right] \geq x_0 - b \cdot \frac{x_0}{b} = 0$$

and

$$x_0 - bt = x_0 - b \cdot \left[\frac{x_0}{b} \right] < x_0 - b \cdot \frac{x_0 - 1}{b} = b$$

Hence we have $0 \leq x \leq b - 1$. It is followed by:

$$
\begin{aligned}
by &= n - ax & &\because ax + by = n \\
&\geq (1 + ab - a - b) - ax & &\because n > ab - a - b \\
&\geq (1 + ab - a - b) - a(b - 1) & &\because x \leq b - 1 \\
&= 1 - b
\end{aligned}
$$

Because b is a positive integer, the above inequality implies:

$$
y \geq \frac{1 - b}{b} = \frac{1}{b} - 1 \geq 0
$$

Hence, such (x, y) is a non-negative solution. This implies the equation is solvable in non-negative integers.

Proof of statement (ii)

From $ax + by = n$ and $n = ab - a - b$, we have

$$
ax + by = ab - a - b \implies a(x + 1) + b(y + 1) = ab
$$

This means that if $ax + by = ab - a - b$ has non-negative integer solution, then the equation $aX + bY = ab$ is solvable in positive integers. We now show this is impossible.

Let (X, Y) is such a solution, then

$$
aX + bY \equiv ab \equiv 0 \pmod{a} \implies bY \equiv 0 \pmod{a}
$$

and

$$
aX + bY \equiv ab \equiv 0 \pmod{b} \implies aX \equiv 0 \pmod{b}
$$

Because $\gcd(a, b) = 1$, the above two implies:

$$
\begin{cases}
Y \equiv 0 & \pmod{a} \\
X \equiv 0 & \pmod{b}
\end{cases}
$$

Consequently, we have $ab = aX + bY \geq ab + ba = 2ab$ which cannot hold!

C.7 Pythagorean Triplets

Practice 1

In the primitive Pythagorean formula below, what are the properties of x, y, z, m and n?

$$\begin{cases} x & = m^2 - n^2 \\ y & = 2mn \\ z & = m^2 + n^2 \end{cases}$$

They must satisfy the following conditions:

(i) y is even, x and z are odd

(ii) $m > n$, $\gcd(m, n) = 1$

(iii) m and n are of opposite parties

Practice 2

Let integers a, b and c be the lengths of a Pythagorean triangle's three sides, where $c > a, b$. Show that

$$\frac{1}{2}(c - a)(c - b)$$

must be a square number.

By the Pythagorean triplets formula, let

$$\begin{cases} a = m^2 - n^2 \\ b = 2mn \\ c = m^2 + n^2 \end{cases}$$

Then

$$\frac{1}{2}(c-a)(c-b)$$
$$=\frac{1}{2}((m^2+n^2)-(m^2-n^2))((m^2+n^2)-2mn)$$
$$=\frac{1}{2}\cdot 2n^2\cdot(m-n)^2$$
$$=n^2(m-n)^2$$

which clearly is a square number.

Practice 3

Show that a Pythagorean triangle must:

(i) have least one side whose length is a multiple of 3, and

(ii) have least one side whose length is a multiple of 4, and

(iii) have least one side whose length is a multiple of 5

Note that these sides may not be necessarily distinct. For example, in a 5-12-13 triangle, the side 12 is a multiple of both 3 and 4.

By the Pythagorean triplet formula, the lengths of the three sides can be expressed as:

$$\begin{cases} x = m^2 - n^2 \\ y = 2mn \\ z = m^2 + n^2 \end{cases}$$

where $m > n$ are two positive integers.

Proof: the length of one side must be a multiple of 3

If either m or n is a multiple of 3, then $y = 2mn$ must be a multiple of 3.

Otherwise, if neither m nor n is a multiple of 3, then $x = m^2 - n^2$ must be a multiple of 3. (See *Example A.3.5* on *page 107.*)

Proof: the length of one side must be a multiple of 4

If either m or n is an even number, then $y = 2mn$ must be a multiple of 4.

Otherwise, if neither m nor n is an even number, then $x = m^2 - n^2$ must be a multiple of 4 because it is the difference of two squares of odd numbers. Please refer to the discussion of *Example A.3.1* on *page 105.*

Proof: the length of one side must be a multiple of 5

If either m or n is a multiple of 5, then $y = 2mn$ must be a multiple of 5.

Otherwise, if neither m nor n is a multiple of 5, then by the conclusion of *Example A.3.6* on *page 108*, either $x = m^2 - n^2$ or $z = m^2 + n^2$ is a multiple of 5.

Practice 4

How many grid points whose x and y coordinates are both integers locate on the circle centered at $(199, 0)$ with a radius of 199?

This question is equivalent to determining the number of integers solutions to the equation:

$$(x - 199)^2 + y^2 = 199^2$$

Clearly, $(0,0)$, $(398,0)$, $(199,199)$, and $(199,-199)$ are four qualified solutions.

If $x \neq 199$, and $y \neq 0$, then $|x - 199|$, y, and 199 form a primitive Pythagorean triplet because, as a prime number, 199 must be relatively prime with both y and $|x - 199|$.

By the Pythagorean triplet formula, let $199 = m^2 + n^2$. However this cannot hold because

$$199 \equiv 3 \pmod 4$$

but the sum of two squares cannot have this property.[1]

Therefore, there are just four qualified grid points on the said circle.

---

[1]Please refer to *Example A.3.7* on *page 108*.

C.8 The Infinite Descent Method

Practice 1

Show that if a positive integer n is not a square number, then $\sqrt{n}$ is an irrational number.

Because n is not a square number, let $n = kd^2$ where k is not divisible by any square number greater than 1. Then the original claim is equivalent to show $\sqrt{k}$ is an irrational number. If this is not the case, let $\sqrt{d} = \frac{p}{q}$ where p and q are positive integers that minimize the value of $(p + q)$.

It follows that $kq^2 = p^2$. Because k is not divisible by any square number, it must hold that $k \mid p$.

Setting $p = kp_1$ leads to $kq^2 = (kp_1)^2$, or $q^2 = kp_1^2$. By the same argument, we find $k \mid q$.

Letting $q = kq_1$ yields $(kq_1)^2 = kp_1^2$, or $kq_1^2 = p_1^2 \implies \sqrt{k} = \frac{p_1}{q_1}$ where both p_1 and q_1 are positive integers.

As positive integer $k \neq 1$, we have $p_1 + q_1 = \frac{p}{k} + \frac{q}{k} < p + q$. This contradicts to the minimality of $(p + q)$. Therefore we conclude $\sqrt{k}$ must be irrational. This is followed by the original claim that $\sqrt{n}$ is irrational.

Practice 2

Show that the area of a Pythagorean triangle cannot be a square number.

It is sufficient to show that there is no such primitive Pythagorean triangle exists. This is because any non-primitive Pythagorean triangle's area is a square number times of that of the corresponding primitive Pythagorean triangle.

If there is such a primitive Pythagorean triangle, its three sides can be expressed as

$$\begin{cases} x &= m^2 - n^2 \\ y &= 2mn \\ z &= m^2 + n^2 \end{cases}$$

where positive integers m and n are relatively prime and are of opposite parity.

Accordingly, its area is $S = \frac{1}{2}(2mn)(m^2 - n^2) = mn(m^2 - n^2)$.

Because m and n are relatively prime, it is easy to show that both $\gcd(m, m^2 - n^2) = 1$ and $\gcd(n, m^2 - n^2) = 1$ hold. Hence, if S is a square number, all of m, n and $(m^2 - n^2)$ must be square numbers. Let

$$\begin{cases} m &= u^2 \\ n &= v^2 \\ m^2 - n^2 &= w^2 \end{cases} \tag{C.13}$$

Observe that we now have a new Pythagorean triplet (w, n, m). Because $\gcd(m, m^2 - n^2) = 1$ and $\gcd(n, m^2 - n^2)$, this triplet is a primitive triplet. Therefore, w and m are odd, and n is even, by properties of the Pythagorean formula. This follows that v is also even.

Because (w, n, m) is a Pythagorean triplet, there exist positive integers a and b such that

$$\begin{cases} w &= a^2 - b^2 \\ n &= 2ab \\ m &= a^2 + b^2 \end{cases}$$

Now, we have another Pythagorean triplet $a^2 + b^2 = m = u^2$. The 2nd equality derives from *Equation C.13*. Therefore, the area of this triangle is

$$\frac{1}{2}ab = \frac{n}{4} = \left(\frac{v}{2}\right)^2$$

Note v is even, therefore this area is square.

Therefore we construct a smaller triangle (a, b, u) from the triangle (x, y, z) that also satisfies the condition. This construction process is repeatable and thus is impossible by the principle of infinite descent.

Practice 3

Show that $x^4 + y^4 = z^2$ is not solvable in positive integers.

If this equation has positive integer solution, let (x_0, y_0, z_0) be the solution that minimizes the value of z.

First, let's show that x_0 and y_0 are relatively prime. If this is not true, let $\gcd(x_0, y_0) = d > 1$. It follows that $d^4 \mid x_0^4 + y_0^4 = z_0^2$. Hence $d^2 \mid z_0$. Therefore $(\frac{x_0}{d}, \frac{y_0}{d}, \frac{z_0}{d^2})$ is another positive integer solution with smaller z. This is a contradiction.

If (x_0, y_0) are relative prime, then (x_0^2, y_0^2, z_0) forms a primitive Pythagorean triplet. Let

$$\begin{cases} x_0^2 = m^2 - n^2 \\ y_0^2 = 2mn \\ z_0 = m^2 + n^2 \end{cases} \qquad \text{(C.14)}$$

where m and n are two relatively prime positive integers. Furthermore, x^2 is odd, thus x is odd.

From this, we have another Pythagorean triplet: $x^2 + n^2 = m^2$. Hence, there exist positive integers a and b such that (note x is odd):

$$\begin{cases} x = a^2 - b^2 \\ n = 2ab \\ m = a^2 + b^2 \end{cases} \qquad \text{(C.15)}$$

Combining *Equation C.14* and *C.15* leads to

$$y^2 = 2mn = 4ab(a^2 + b^2)$$

Because a and b are relatively prime, it can be shown that ab and $(a^2 + b^2)$ are relatively prime too. Consequently, all of a, b, and $(a^2 + b^2)$ are squares. That is, there exits some integers u, v, and w such that

$$\begin{cases} a & = u^2 \\ b & = v^2 \\ a^2 + b^2 & = w^2 \end{cases}$$

However, this means we found another solution (u, v, w) satisfies the relationship:

$$u^4 + v^4 = w^2$$

It is also clear that $w < z_0$ which violates the minimality of z_0. Hence, by the principle of infinite descent, this is impossible.

Practice 4

Show that when $n = 4$, the Fermat's equation

$$x^n + y^n = z^n$$

is not solvable in positive integers.

This is a natural extension of the previous practice.

Because $x^4 + y^4 = z^2$ is unsolvable in positive integers, the equation

$$x^4 + y^4 = (z^2)^2$$

is unsolvable in positive integers either.

Practice 5

Prove that if positive integers a and b are such that $ab+1$ divides $a^2 + b^2$. then

$$\frac{a^2 + b^2}{ab + 1}$$

is a square number.

(1988 IMO)

Let integer k satisfy $k = \frac{a^2+b^2}{ab+1}$, or

$$a^2 - kab + b^2 - k = 0 \qquad \text{(C.16)}$$

The goal is to prove k is a square.

Let's assume the positive integer pair (a_0, b_0) satisfies the above relationship which minimizes the value of $(a + b)$. By symmetry, let's also assume $a_0 \geq b_0 > 0$.

It is clear from *Equation C.16* that a_0 is a root of the quadratic equation

$$x^2 - kb \cdot x + (b^2 - k) = 0$$

Assume the other root of this equation is a_1. By the Vieta's theorem:

$$\begin{cases} a_0 + a_1 &= k \cdot b \\ a_0 \cdot a_1 &= b^2 - k \end{cases} \qquad \text{(C.17)}$$

The 1st equation above implies a_1 is an integer.

If k is not a square, then the 2nd equation above implies $a_1 \neq 0$. This is because otherwise k will be a square number. However this will lead to the conclusion that $0 < a_1 < a_0$ (see proof below). This implies there exists another smaller positive integer pair (a_1, b_0) satisfying *Equation C.16*. This contradicts the minimality assumption of (a_0, b_0).

First, a_1 must be positive. Otherwise, the following will hold (because b_0 is a positive integers):

$$a_1^2 - ka_1b_0 + b_0^2 - k \geq a_1^2 + k + b_0^2 - k = a_1^2 + b_0^2 > 0$$

This contradicts the fact that a_1 is a solution to *Equation C.17*.

Meanwhile, by *Equation C.17* and the assumption $a_0 \geq b_0$, the following must hold:

$$a_1 = \frac{b_0^2 - k}{a_0} \leq \frac{b_0^2 - 1}{a_0} \leq \frac{a_0^2 - 1}{a_0} < a_0$$

C.9   Pell's Equation

Practice 1

Why is it usually assumed that d is not a square number when discussing the Pell's equation $x^2 - dy^2 = 1$?

This is because when d is a square number, the equation only has trivial solutions $(\pm 1, 0)$.

Practice 2

Show that if integer pair (x, y) satisfies $x^2 - 2y^2 = -1$, then both x and y are odd.

It is obvious that x must be odd. Let it be $x = (2k + 1)$. Then we have:

$$(2k + 1)^2 - 2y^2 = -1$$
$$4k^2 + 4k + 1 - 2y^2 = -1$$
$$y^2 = 2k^2 + 2k + 1$$

It is clearly now that y^2 must be odd which is followed by the claim that y is odd too.

Practice 3

Find all positive integer solutions to the equation:

$$x^2 - 5y^2 = 1$$

Its fundamental solution is $(9, 4)$. Hence, its general solution is:

$$\begin{cases} x_n & = \dfrac{1}{2}((9 + 4\sqrt{5})^n + (9 - 4\sqrt{5})^n) \\ \\ y_n & = \dfrac{1}{2\sqrt{5}}((9 + 4\sqrt{5})^n - (9 - 4\sqrt{5})^n) \end{cases}$$

where n is a positive integer.

Practice 4

Show that if x and y are positive integer solutions to the equation $x^2 - 2y^2 = 1$, then $6 \mid xy$.

Its fundamental solution is $(x_1, y_1) = (3, 2)$. Therefore all solutions satisfy

$$\begin{cases} x_{n+1} & = 3x_n + 4y_n \\ \\ y_{n+1} & = 3y_n + 2x_n \end{cases}$$

When $n = 1$, $x_1 y_1 = 3 \times 2 = 6$. The claim holds.

Let's assume the claim holds when $n = k$, i.e. $6 \mid x_k y_k$.

When $n = k + 1$, we have

$$x_{k+1} y_{k+1} = (3x_k + 4y_k)(3y_k + 2x_k) = 6x_k^2 + 12y_k^2 + 27x_k y_k$$

Clearly, every term on the right is a multiple of 6. Hence the claim holds too.

By the principle of mathematical induction, the claim holds for all the positive integer solutions.

Practice 5

Solve in integers the equation $x^2 + y^2 - 1 = 4xy$.

The given equation is equivalent to $(x - 2y)^2 - 3y^2 = 1$.

Substituting $u = x - 2y$ leads to the following Pell's equation:

$$u^2 - 3y^2 = 1$$

Its fundamental solution is $(u_1, y_1) = (2, 1)$. Hence all its solutions are given by:

$$\begin{cases} u_n & = & \frac{(2+\sqrt{3})^n + (2-\sqrt{3})^n}{2} \\ \\ y_n & = & \frac{(2+\sqrt{3})^n - (2-\sqrt{3})^n}{2\sqrt{3}} \end{cases}$$

Accordingly the solutions to the original equations are given by:

$$\begin{cases} x_n & = & \frac{(2+\sqrt{3})^n + (2-\sqrt{3})^n}{2} + \frac{(2+\sqrt{3})^n - (2-\sqrt{3})^n}{\sqrt{3}} \\ \\ y_n & = & \frac{(2+\sqrt{3})^n - (2-\sqrt{3})^n}{2\sqrt{3}} \end{cases}$$

Because the given equation is symmetrical, therefore for every solution (x_n, y_n), (y_n, x_n) is also a solution.

Practice 6

Solve in *rational* numbers the equation $x^2 - dy^2 = 1$ where d is an integer.

No knowledge of Pell's equation is required when the desired solutions are rational numbers.

Rearranging the original equation to the following:

$$x^2 - 1 = dy^2$$
$$(x+1)(x-1) = dy^2$$
$$\frac{x+1}{y} = d\left(\frac{y}{x-1}\right)$$

Setting parameter $t = \frac{y}{x-1} \in \mathbb{Q}$:

$$\frac{x-1}{y} + \frac{2}{y} = dt \implies y = \frac{2t}{dt^2 - 1}$$

Accordingly,

$$x = \frac{dt^2 + 1}{dt^2 - 1}$$

Practice 7

Find all positive integers k, m such that $k < m$ and

$$1 + 2 + \cdots + k = (k+1) + (k+2) + \cdots + m$$

Adding $1 + 2 + \cdots + k$ to both sides leads to:

$$2 \times \frac{k(k+1)}{2} = \frac{m(m+1)}{2}$$

This can be rewritten as the following Pell's equation

$$(2m+1)^2 - 2(2k+1)^2 = -1$$

Now for equation $x^2 - 2y^2 = -1$, its fundamental solution is $(1, 1)$. Hence it has infinitely many solutions in the following form:

$$\begin{cases} x_n = \dfrac{(1+\sqrt{2})^{2n-1} + (1-\sqrt{2})^{2n-1}}{2} \\[4mm] y_n = \dfrac{(1+\sqrt{2})^{2n-1} - (1-\sqrt{2})^{2n-1}}{2\sqrt{2}} \end{cases}$$

It follows that $(k, m) = \left(\dfrac{y_n - 1}{2}, \dfrac{x_n - 1}{2} \right)$.

By *Practice C.9* on *page 153*, solutions (x_n, y_n) to $x^2 - 2y^2 = -1$ are both odd. Therefore (k, m) are both integers. Meanwhile, $k < m$ obviously holds.

Practice 8

Let r be a positive real number, and $[r]$ be the largest integer not exceeding r. Prove that there exist unlimited number of positive integers, n, such that $[\sqrt{2}\, n]$ is a perfect square.

For a positive integer n, if we can find a positive integer, x, such that

$$x^2 < \sqrt{2}\, n < x^2 + 1 \tag{C.18}$$

then $[\sqrt{2}\, n] = x^2$. If there exist infinitely many such n, then the claim is proved.

Consider the equation

$$x^2 - 2y^2 = -1 \tag{C.19}$$

Multiplying x^2 on both sides of the equation and rearranging the terms leads to: $2x^2 y^2 = x^4 + x^2$.

Therefore: $x^4 < 2x^2 y^2 < (x^2 + 1)^2$, or $x^2 < \sqrt{2}\, xy < x^2 + 1$.

Let $n = xy$, we then have $x^2 < \sqrt{2}\, n < (x + 1)^2$.

Because *Equation C.19* has one solution $(1, 1)$, by *Theorem 9.3.4*, it has infinite number of solutions. In another word, there exist infinitely many pairs of integers (x, n) satisfying the relationship *C.18*. Hence the claim holds.

C.10   The MOD Method

Practice 1

Solve the following equation in integers: $x^2 + y^2 = 2015$.

It is unsolvable because $x^2 + y^2 \equiv 2015 \equiv 3 \pmod 4$ will never hold if both x and y are integers.

Practice 2

Prove: $2x^2 - 5y^2 = 7$ has no integer solution.

Clearly, $5y^2$ must be an odd number, which means y is odd too. It follows that

$$y^2 \equiv 1 \pmod 4 \qquad \text{and} \qquad y^2 \equiv 1 \pmod 8$$

Now we are going to show that there will be a contradiction regardless of x's parity.

If x is even, then take $\pmod 8$ on both sides:

$$2x^2 - 5y^2 \equiv 7 \pmod 8$$
$$0 - 5 \times 1 \equiv 7 \pmod 8$$
$$3 \equiv 7 \pmod 8$$

Obviously, this cannot be true.

If x is odd, then take $\pmod 4$ on both sides:

$$2x^2 - 5y^2 \equiv 7 \pmod 4$$
$$2 \times 1 - 5 \times 1 \equiv 3 \pmod 4$$
$$1 \equiv 3 \pmod 4$$

It will not hold either.

Therefore, we conclude that the original equation has no integer solution.

Practice 3

Find all positive integer solutions to the equation $3^x - 2^y = 1$.

Clearly, $x = y = 1$ is one solution.

If $y \geq 2$, then 2^y must be a multiple of 4. Taking (mod 4) on both sides yields:

$$(-1)^x - 0 \equiv 1 \quad (\text{mod } 4)$$

This implies x is even. Setting $x = 2m$ yields

$$3^{2m} - 2^y = 1$$

Rearranging the equation, and then applying the difference of squares formula:

$$2^y = (3^m + 1)(3^m - 1) \tag{C.20}$$

Its left side is some power of 2. Therefore both terms on the right side must be some power of 2 as well. Let

$$\begin{cases} 3^m + 1 &= 2^t \\ 3^m - 1 &= 2^{y-t} \end{cases} \tag{C.21}$$

where t is an integer and $3^m + 1 > 3^m - 1 \implies t > y - t$.

Subtracting the second equation from the first one leads to:

$$2 = 2^t - 2^{y-t} = 2^{y-t}(2^{2t-y} - 1) \tag{C.22}$$

Because there is only one way to factorize the number 2, the following must hold:

$$\begin{cases} 2^{y-t} &= 2 \\ 2^{2t-y} - 1 &= 1 \end{cases}$$

Now everything flows:

$$2^{y-t} = 2^{2t-y} = 2 \implies (y,t) = (3,2)$$

Setting $y = 3$ to the original equation leads to $x = 2$.

In conclusion, there are two positive integer solutions:

$$(1,1) \quad \text{and} \quad (2,3)$$

Practice 4

Solve this equation in positive integers: $|12^x - 5^y| = 7$.

Casework is the best to tackle equations involving absolute values.

Case 1: if $12^x < 5^y$, then $5^y - 12^x = 7$

Take (mod 4) on both sides:

$$
\begin{aligned}
5^y - 12^x &\equiv 7 && (\text{mod } 4) \\
1^y - 0 &\equiv 3 && (\text{mod } 4) \\
1 &\equiv 3 && (\text{mod } 4)
\end{aligned}
$$

Clearly, it cannot hold which means no solution under this circumstance.

Case 2: if $12^x > 5^y$, then $12^x - 5^y = 7$

Take (mod 3) on both sides:

$$
\begin{aligned}
12^x - 5^y &\equiv 7 && (\text{mod } 3) \\
0 - (-1)^y &\equiv 1 && (\text{mod } 3) \\
(-1)^y &\equiv -1 && (\text{mod } 3)
\end{aligned}
$$

Therefore y must be odd. Letting $y = 2k + 1$ yields

$$12^x - 5^{2k+1} = 7$$
$$12^x - 5 \times 25^k = 7$$

Then, take (mod 8) on both sides yields:

$$12^x - 5 \times 25^k \equiv 7 \qquad (\text{mod } 8)$$
$$4^x - 5 \times 1^k \equiv 7 \qquad (\text{mod } 8)$$
$$4^x \equiv 12 \qquad (\text{mod } 8)$$
$$4^x \equiv 4 \qquad (\text{mod } 8)$$

The last equation implies $4^x = 8p + 4$ where p is an integer. Dividing 4 on both sides:

$$4^{x-1} = 2p + 1$$

The right side of this equation is odd. The only possibility for the left side to be an odd number is $x = 1$. Setting $x = 1$ in the original equation $12^x - 7^y = 1$ leads to $y = 1$.

Therefore, this equation has only one positive integer solution:

$$(x, y) = (1, 1)$$

Practice 5

Find all positive integer solutions to the equation

$$8^x + 15^y = 17^z$$

Taking (mod 8) on both sides:

$$(-1)^y \equiv 1^z \quad (\text{mod } 8) \implies y \text{ must be even}$$

Taking (mod 7) on both sides:

$$1^x + 1^y \equiv 3^z \quad (\text{mod } 7) \implies z \text{ is even too}$$

Letting $y = 2m$ and $z = 2n$ and rearranging the equation lead to

$$8^x = 17^{2n} - 15^{2m}$$
$$2^{3x} = (17^n - 15^m) \times (17^n + 15^m)$$

We find that both $(17^n - 15^m)$ and $(17^n + 15^m)$ are some power of 2. Let's assume:

$$\begin{cases} 17^n - 15^m = & 2^t \\ 17^n + 15^m = & 2^{3x-t} \end{cases} \tag{C.23}$$

Solving these two equations with respect to 17^n and 15^m yields:

$$17^n = \frac{1}{2} \times (2^t + 2^{3x-t})$$
$$2 \times 17^n = 2 \times (2^{t-1} + 2^{3x-t-1})$$

Because 17 is odd, $2^{t-1} + 2^{3x-t-1}$ must be odd, either 2^{t-1} or 2^{3x-t-1} must equal 1. It follows that one of the following must hold:

$$t - 1 = 0 \quad \text{or} \quad 3x - t - 1 = 0$$

If $t - 1 = 0$, then from *Equation C.23*, we have $17^n - 15^m = 2$. In this case, $m = n = 1$ is the only solution[2]. We find $y = z = 2$, and finally $x = 2$.

If $3x - t - 1 = 0$, then $17^n + 15^m = 2$ which is impossible.

As such, we conclude $(2, 2, 2)$ is the only solution.

Practice 6

Solve in integers the equation $|3^x - 2^y| = 41$.

It is natural to do casework when absolute value is involved in an equation.

Case 1: $3^x - 2^y = 41$

Take $(\bmod 3)$ on both sides: $0 - (-1)^y \equiv -1 \pmod 3$. Therefore y must be even. Let it be $2n$.

---

[2]Please refer to *Example 10.2.2* on *page 93*.

Take (mod 4) on both sides: $(-1)^x - 0 \equiv 1$ (mod 4). Therefore x is even too. Let it be $2m$.

Therefore by the difference of squares formula:

$$3^{2m} - 2^{2n} = 41$$
$$(3^m + 2^n)(3^m - 2^n) = 41$$

Because 41 is prime, the following must hold:

$$\begin{cases} 3^m + 2^n = 41 \\ 3^m - 2^n = 1 \end{cases}$$

Adding these two equations yields $2 \times 3^m = 42$. This is clearly insolvable in integers.

Case 2: $2^y - 3^x = 41$

Obviously, it must hold that $y > 3$. Therefore $8 \mid 2^y$. Taking (mod 8) on both sides leads to $0 - 3^x \equiv -7$ (mod 8), or

$$3^x \equiv 7 \pmod 8$$

However, this cannot hold because, for any given positive integer, it can be shown that $3^x \equiv 1$ or $3^x \equiv 3$ (mod 8).

Hence, we find that the given equation has no integer solution.

Practice 7

Find all positive integer solutions to $7^x - 3 \times 2^y = 1$.

It is obvious that $x = y = 1$ is one solution.

When $x \geq 2$, we must have $y \geq 4$. Take (mod 8) on both sides:

$$(-1)^x - 0 \equiv 1 \pmod 8$$

Therefore x must be even. Let it be $2k$:

$$7^{2k} - 3 \times 2^y = 1$$
$$7^{2k} - 1 = 3 \times 2^y$$
$$(7^k + 1)(7^k - 1) = 3 \times 2^y$$

We find exact one of $(7^k - 1)$ and $(7^k + 1)$ is a multiple of 3. Hence, one of the following systems must hold:

(A) $\begin{cases} 7^k + 1 = 3 \times 2^m \\ 7^k - 1 = 2^{y-m} \end{cases}$ or (B) $\begin{cases} 7^k + 1 = 2^m \\ 7^k - 1 = 3 \times 2^{y-m} \end{cases}$

where m is an integer and satisfies $0 \le m \le y$.

System (A) is impossible because taking (mod 3) on the first equation will result in an impossible relationship:

$$1^k + 1 \equiv 0 \pmod 3$$

System (B) has one obvious solution: $(k, m) = (1, 3)$. Accordingly, $(x, y) = (2, 4)$. We now show there is no solution for $k > 1$.

Because $7^k + 1 = 2^m$, we must have $m \ge 3$ which means $8 \mid 2^m$.

Taking (mod 8) on both sides of $7^k + 1 = 2^m$ leads to:

$$(-1)^k + 1 \equiv 0 \pmod 8$$

Therefore k must be odd. It follows that:

$$2^m = 7^k + 1 = (7 + 1) \times (7^{k-1} - 7^{k-2} + \cdots - 7 + 1)$$

The 2nd bracket on the right side has k terms, and all the terms are odd. Because integer k is odd, this sum must be odd. Furthermore, because $k > 1$, this sum must be greater than 1.

However the left side is 2^m. It cannot be divisible by any odd number that is greater than 1.

Therefore, in conclusion, there are only two solutions:

$$(1, 1) \text{ and } (2, 4)$$

Practice 8

Solve the following equation in positive integers:

$$2015x + 2000y + 3 = z^2$$

Because $5 \mid 2015x + 2000y$, the left side $2015x + 2000y + 3$ must end with either digit 3 or 8. However no square number will end with these two digits.

Consequently, this equation has no integer solution.

www.ingramcontent.com/pod-product-compliance
Lightning Source LLC
Chambersburg PA
CBHW051913170526
45168CB00001B/362